U0865118

HIT

国外优秀数学著作
原　版　系　列

带有潜在故障恢复系统的半马尔柯夫模型控制

Semi-Markov Models Control of Restorable Systems with Latent Failures

[俄罗斯] Obzherin, Y. E.（阿布泽瑞） [俄罗斯] Boyko, E. G.（博伊科） 著

HITP
哈爾濱工業大學出版社
HARBIN INSTITUTE OF TECHNOLOGY PRESS

黑版贸审字 08-2015-063 号

Semi-Markov Models Control of Restorable Systems with Latent Failures
Yuriy E. Obzherin and Elena G. Boyko
ISBN:978-0-12-802212-2
Copyright © 2015 Elsevier Inc. All rights reserved.

Authorized English language reprint edition published by Elsevier (Singapore) Pte Ltd. and Harbin Institute of Technology Press

Copyright © 2016 by Elsevier (Singapore) Pte Ltd. All rights reserved.
Elsevier (Singapore) Pte Ltd.
3 Killiney Road, #08-01 Winsland House I, Singapore 239519
Tel: (65)6349-0200 Fax: (65) 6733-1817
First Published 2016
2016 年初版
Printed in China by Harbin Institute of Technology Press under special arrangement with Elsevier (Singapore) Pte Ltd. This edition is authorized for sale in China only, excluding Hong Kong SAR, Macao SAR and Taiwan. Unauthorized export of this edition is a violation of the Copyright Act. Violation of this Law is subject to Civil and Criminal Penalties.

本书英文影印版由 Elsevier (Singapore) Pte Ltd. 授权哈尔滨工业大学出版社在中国大陆境内独家发行。本版仅限在中国境内(不包括香港、澳门以及台湾)出版及标价销售。未经许可之出口,视为违反著作权法,将受民事及刑事法律之制裁。
本书封底贴有 Elsevier 防伪标签,无标签者不得销售。

图书在版编目(CIP)数据

带有潜在故障恢复系统的半马尔柯夫模型控制:英文/(俄罗斯)阿布泽瑞(Obzherin, Y. E.),(俄罗斯)博伊科(Boyko, E. G.)著. —哈尔滨:哈尔滨工业大学出版社,2016.1
书名原文:Semi-Markov Models Control of Restorable Systems with Latent Failures
ISBN 978-7-5603-5760-7

Ⅰ.①带… Ⅱ.①阿…②博… Ⅲ.①半马尔柯夫过程-英文 Ⅳ.①O211.62

中国版本图书馆 CIP 数据核字(2015)第 292131 号

策划编辑 刘培杰
责任编辑 张永芹 钱辰琛
封面设计 孙茵艾
出版发行 哈尔滨工业大学出版社
社　　址 哈尔滨市南岗区复华四道街 10 号 邮编 150006
传　　真 0451-86414749
网　　址 http://hitpress.hit.edu.cn
印　　刷 哈尔滨市工大节能印刷厂
开　　本 787mm×1092mm 1/16 印张 14.25 字数 282 千字
版　　次 2016 年 1 月第 1 版 2016 年 1 月第 1 次印刷
书　　号 ISBN 978-7-5603-5760-7
定　　价 98.00 元

(如因印装质量问题影响阅读,我社负责调换)

Contents

Preface

The improvement of industrial systems' reliability and production quality is a real problem of the modern industry. The automatic systems of technological processes management enable solving this problem. The local technical control constitutes an important part of the system.

In spite of the diversity and high level of checkout and measurable instruments, the problem of latent failures detection and elimination is still significant.

The latent failure is a failure which cannot be detected by standard methods or visually, but by maintenance or special methods of diagnostics only.

In this monograph, by the latent failure we refer to one which can be detected during control execution only.

In complex industrial systems, the periodical control is applied. The reason is the difficulty of checking individual operation of all units and details (components). It means the control is carried out at fixed (in general case random) time periods, which should be optimal for the whole system by ensuring its maximum reliability and efficiency. The problem can be solved by constructing mathematical models of control of restorable systems with latent failures.

The present monograph is dedicated to building such models on the basis of the theory of semi-Markov processes with arbitrary phase state space, as well as to the definition of optimal periodicity of latent failures control. The problems of application of the results obtained are considered.

The authors express deep gratitude to professor V.J. Kopp and professor A.I. Peschansky for valuable comments, contributing to the monograph quality improvement, and to A.I. Kovalenko for the monograph translation into English.

List of Notations and Abbreviations

ADS CPM	automatic decision system of control periodicity management
DD	distribution density
DF	distribution function
EMC	embedded Markov chain
MRP	Markov renewal process
RT	restoration time
RV	random variable
SM	semi-Markov
SMP	semi-Markov process
TF	time to failure
$\alpha \wedge \beta =$	$\min(\alpha, \beta) = \begin{cases} \alpha, & \alpha < \beta, \\ \beta, & \alpha \geq \beta \end{cases}$
$dx = (x, x + dx)$	differential of numeric variable x, denotes interval and interval length
E	system phase state space
E_+, E_-	sets of up- and down-system states, respectively
$E^{(0)}$	set of ergodic system states
K_a	stationary availability factor
$E\alpha$	expectation of random variable α
P(A)	probability of A event
T_+	average stationary operating time to failure
T_-	average stationary restoration time
S	average specific income per calendar time unit
C	average specific expenses per time unit of up-state

Introduction

Technical control is an important part of the production quality control department of any enterprise. The rapid development of technologies, increase in quality, and reliability requirements result in considerable growth of technical control expenses. As noted in [6], metal-processing industry spends 8–15% of expenses on the quality control. It takes from 5 hours to several weeks to make the project of a single detail control, and from 40 minutes to several hours to execute its control. That is why the tasks of reduction of control expenses and efficiency increase are significant.

Mathematical models of technical control execution can serve to solve the problem. These models allow one to analyze the efficiency of different control strategies and to define optimal periods of their execution. The present monograph is dedicated to the control modeling with regard to latent failures of the technical system.

According to the possibility of detection, there are two types of failures [20]:

- evident failure, which can be detected visually or by standard methods of control and diagnostics in the process of object preparation and exploitation;
- latent failure, which can be detected by maintenance or special methods of diagnostics only.

A great number of parametric failures are referred to as latent.

As stated in the Preface, by the latent failure we mean the one which can be detected in the control process.

In the present monograph, to build control models, the approach introduced by V.S. Korolyuk, A.F. Turbin, and their disciples [13–17] is used. It is based on the application of the theory of semi-Markov processes with arbitrary phase space. This approach allows us to omit some restrictions, in particular the assumption of exponential distribution laws of random variables, describing the system. It enables obtaining applicable system operation characteristics. In cases of high model dimensions, algorithms of phase merging serve as an efficient approximation method [14–17].

In this present monograph, the concept of a system component is involved. A component is a constituent part or element of a system. If a system functionally consists of one element (component), not divisible from the point of view of failures, it is called a one-component. The system consisting of $n \geq 2$ indivisible components is named multicomponent [4, 20].

In the present work one- and two-component restorable systems with latent failures control are investigated. However, the approach can be applied to multicomponent systems [21, 22].

In Chapter 1 of the monograph, preliminaries are given.

Chapter 2 covers semi-Markov models for different control strategies in one-component systems. Their stationary characteristics of reliability and efficiency are defined. For the characteristics approximation, we apply the method offered in [14]. It has common background with algorithms of asymptotic phase merging.

Chapter 3 is dedicated to semi-Markov models of latent failures control in two-component systems.

In Chapter 4, on the basis of the results obtained in Chapters 2 and 3, the problems of optimal periodicity of control execution are solved.

Chapter 5 contains comparative analysis of analytical and imitational modeling of some one- and two-component systems. The possibility of practical application of the results represented in the present monograph is considered.

In Chapter 6 semi-Markov models of systems of different function are considered:

- model of queuing system with losses;
- model of system with a cumulative reserve of time;
- model of two-phase system with a intermediate buffer;
- model of technological cell with nondepreciatory failures.

Appendices include data, to support the reader's understanding of the basic text.

Chapter 1

Preliminaries

Chapter Outline

1.1 STRATEGIES AND CHARACTERISTICS OF TECHNICAL CONTROL

Automatic checkout systems consist of the object, engineering devices, programs, and operator, which enable to carry out automatic control. Control strategy usually means the rule defining the choice of checkout means with regard to the system controlled. There exist efficiency control and preventive control [6]. Efficiency control is checkout of the production capability to fulfil functions under parameters, determined by manuals. Preventive control is a technical checkout for detection and prevention of defects or flaws.

In the monograph, efficiency control is investigated. Efficiency control is devided into ideal and nonideal [4]. Under ideal efficiency control, all the failures are detected immediately and reliably [4]. Under nonideal efficiency control, latent failures and automatic checkout system failures take place [4].

In Figure 1.1, a general scheme of efficiency control execution with the help of automatic checkout systems is presented. One of the control characteristics is its periodicity. The control periodicity is time period between two successive checkout processes, executed by certain control instrument [6].

According to the object, continuous, periodical, and casual kinds of control are singled out. Under continuous control, the information on parameters is received constantly, while under periodical control it happens at certain time intervals. Casual control is carried out at random time intervals [6]. Casual control includes single control. The latter is executed, for instance, before the use of stored system, in case the system reliability is ensured by the storage measures.

In the monograph, periodic control with full efficiency restoration is investigated.

In Sections 2.1, 2.4, 3.3–3.5 and Sections 2.2, 2.3, 3.1, 3.2, system efficiency control with component deactivation and without deactivation while control execution are considered, respectively.

Semi-Markov Models. http://dx.doi.org/10.1016/B978-0-12-802212-2.00001-2
Copyright © 2015 Elsevier Inc. All rights reserved.

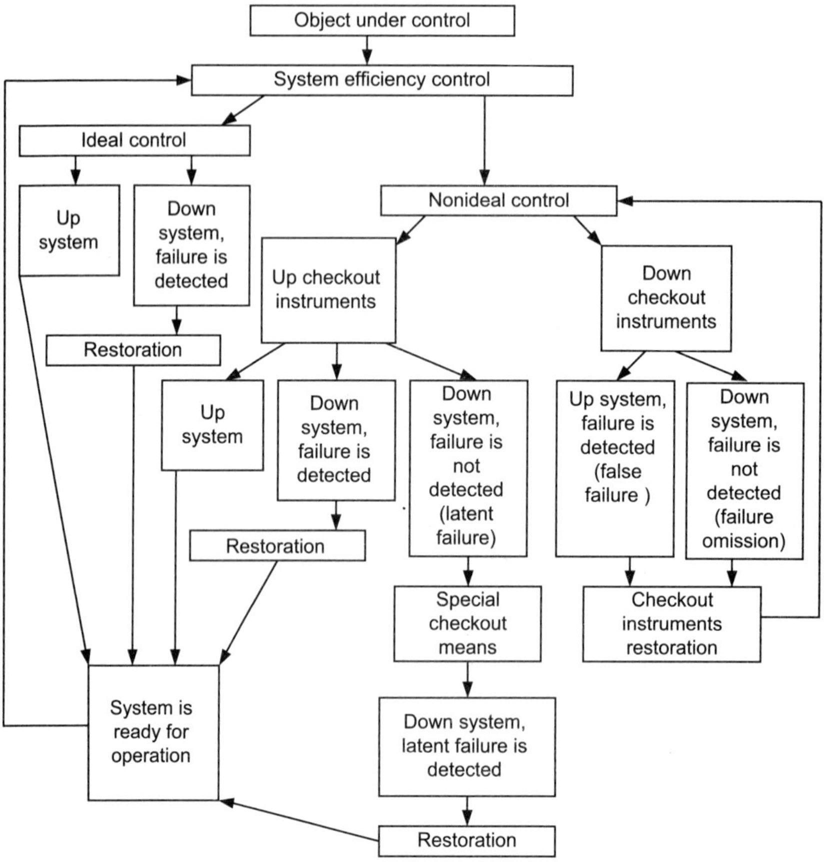

FIGURE 1.1 General scheme of efficiency control execution

1.2 PRELIMINARIES ON RENEWAL THEORY

In the present section, renewal theory review is made [1–3, 7]. The information is mainly given according to monograph [3].

Renewal theory origins from the simplest restoration model: after each failure the system is restored immediately.

Definition [3]. Renewal process is a sequence $\{\alpha_n; n \geq 1\}$ of non-negative independent random variables (RVs) with the same distribution function (DF) $F(t)$.

RVs $T_k = \sum_{n=1}^{k} \alpha_n, k \geq 1, T_0 = 0$ are called renewal moments. Renewal process is often defined as a sequence of RV $\{T_k\}$ as well. The counting renewal process, defined with the help of renewal moments, is of particular interest

$$N(t) = \max\{k : T_k \leq t\}.$$

For each moment t, the value of $N(t)$ determines the random number or renewal moments in $[0, t]$. Renewal function $H(t)$, defining the mean of renewal moments in $(0, t]$, plays fundamental role in renewal theory:

$$H(t) = E[N(t)] = \sum_{k=1}^{\infty} kP\{N(t) = k\} = \sum_{k=1}^{\infty} F^{*(k)}(t), \tag{1.1}$$

where $F^{*(k)}(t)$ is a k-fold convolution of DF $F(t)$, $F^{*(1)}(t) = F(t)$.

The function is a solution of the integral renewal equation:

$$H(t) = F(t) + \int_0^t H(t-x)\,dF(x).$$

Renewal function derivative $h(t) = H'(t) = \sum_{k=1}^{\infty} f^{*(k)}(t)$ is called renewal density. It satisfies the following integral equation:

$$h(t) = f(t) + \int_0^t h(t-x) f(x)\,dx, \tag{1.2}$$

where $f(t)$ is the density of DF $F(t)$.

The following formulas are true:

$$\int_0^t \bar{F}(s) h(t-s)\,ds + \bar{F}(t) = 1, \int_0^t H(t-s)\bar{F}(s)\,ds + \int_0^t \bar{F}(s)\,ds = t. \tag{1.3}$$

Under sufficiently small Δt, the value $h(t)\Delta t$ approximately equals the probability of renewal moment appearance in $(t, t+\Delta t]$.

Explicit form of functions $H(t)$ for some renewal processes can be found in [3, 7].

The functions:

$$\tilde{H}(t) = 1 + H(t), \hat{H}(t) = \begin{cases} 1 + H(t), t > 0, \\ 0, t = 0. \end{cases} \tag{1.4}$$

will be applied as well.

The function $\hat{H}(t)$, having a unit jump at $t = 0$, is used to write down expressions in form of Stieltjes integral.

The following theorems, describing asymptotic behavior (under $t \to +\infty$) of the function $H(t)$, take place.

Renewal theorem (elementary) [3]. For any distribution law of $F(t)$

$$\lim_{t \to +\infty} \frac{H(t)}{t} = \frac{1}{\mu}, \tag{1.5}$$

where $\mu = E\alpha$ is expectation of RV α.

Renewal theorem (key) [3]. If $F(t)$ is not an arithmetic distribution, and $g(t)$ is integrable nonincreasing in $(0,+\infty)$ function, then

$$\lim_{t\to+\infty}\int_0^t g(t-x)dH(x)=\frac{1}{\mu}\int_0^\infty g(x)dx. \quad (1.6)$$

The process of direct residual time $V_t=\tau_{N(t)+1}-t, t\geq 0$, V_t being residual time to failure by t, is connected with the renewal process $\{\alpha_n; n\geq 1\}$. V_t is a homogeneous Markov process with phase state $(0,+\infty)$.

DF $V(t,x)=P\{V_t\leq x\}$ of the direct residual time is defined by the formula [3]:

$$V(t,x)=F(t+x)-\int_0^t \bar{F}(t+x-s)dH(s), \quad (1.7)$$

and corresponding distribution density is:

$$v(t,x)=f(t+x)+\int_0^t f(t+x-u)h(u)du. \quad (1.8)$$

The expectation of the direct residual time equals:

$$E(V_t)=E\alpha(1+H(t))-t. \quad (1.9)$$

In the renewal process, the restoration of failed system is considered to be negligible in comparison with operating time. This assumption does not take place in practice. That is why the following system renewal process is considered [3].

For the first time, the system fails in a random time period α_1 and is restored in random time β_1. The restored system operates for α_2 time, then it fails, and is restored in β_2, and so on. Time moments $T_1=\alpha_1, T_2=\alpha_1+\beta_1+\alpha_2,\ldots,$ of the system failure are called failure moments or moments of 0-renewation, and the moments $S_1=\alpha_1+\beta_1$, $S_2=\alpha_1+\beta_1+\alpha_2+\beta_2,\ldots,$ of the restoration end are restoration moments (or 1-renewations).

Definition [3]. If $\{\alpha_n; n\geq 1\}$ and $\{\beta_n; n\geq 1\}$ are two sequences of independent similarly distributed RVs, the sequence $\{(\alpha_n,\beta_n); n\geq 1\}$, as well as $\{(T_n,S_n); n\geq 1\}$, is called an alternating renewal process.

Alternating renewal process can be equivalently given by $\{Z(t), t\geq 0\}$ with the help of

$$Z(t)=\begin{cases}0, \text{if } t\in(T_k,S_k),\\ 1, \text{otherwise.}\end{cases}$$

According to the definition, the process $Z(t)$ determines system states at the moment t: $Z(t)=1$ for system up-state at t, and $Z(t)=0$ for restoration at t.

Let us denote by $N^{(0)}(t)$ the random number of 0-renewations, and by $N^{(1)}(t)$ the random number of 1-renewations in $(0,t)$. $N^{(0)}(t)$ and $N^{(1)}(t)$ are counting renewal processes, generated by DFs $F(t)$ and $(G*F)(t)$ (correspondingly $(F*G)(t)$).

Mean numbers of 0- and 1-renewations in $(0,t)$ are defined by the expressions:

$$H^{(0)}(t)=E\left[N^{(0)}(t)\right]=\sum_{k=1}^{\infty}F*(G*F)^{*(k-1)}(t), \tag{1.10}$$

$$H^{(1)}(t)=E\left[N^{(1)}(t)\right]=\sum_{k=1}^{\infty}(F*G)^{*(k)}(t). \tag{1.11}$$

Their densities are:

$$h^{(0)}(t)=\sum_{k=1}^{\infty}f*(g*f)^{*(k-1)}(t), h^{(1)}(t)=\sum_{k=1}^{\infty}(f*g)^{*(k)}(t). \tag{1.12}$$

Probabilities $P\{Z(t)=i,V_t^{(i)}>x\}, i=0,1$ are of interest. Here $V_t^{(1)}$ denotes residual time to failure, and $V_t^{(0)}$ stands for residual renewal time.

$P\{Z(t)=1,V_t^{(1)}>x\}$ denotes the probability that the system, operating at the moment t, will not fail in $(t,t+x)$, and $P\{Z(t)=0,V_t^{(0)}>x\}$ is the probability that the system, being restored at t, will not be restored in $(t,t+x)$.

These probabilities can be obtained by [3]:

$$\overline{V}^{(0)}(t,x)=P\{Z(t)=0,V_t^{(0)}>x\}=\int_0^t\overline{G}(t+x-s)h^{(0)}(s)\,ds,$$

$$\overline{V}^{(1)}(t,x)=P\{Z(t)=1,V_t^{(1)}>x\}=\overline{F}(x+t)+\int_0^t\overline{F}(t+x-s)h^{(1)}(s)\,ds.$$

Distribution densities of residual operation time to failure and residual renewal time are:

$$v^{(0)}(t,x)=\int_0^t g(t+x-s)h^{(0)}(s)\,ds, v^{(1)}(t,x)=f(x+t)+\int_0^t f(t+x-s)h^{(1)}(s)\,ds. \tag{1.13}$$

1.3 PRELIMINARIES ON SEMI-MARKOV PROCESSES WITH ARBITRARY PHASE SPACE OF STATES

We represent necessary results from the theory of semi-Markov processes (SMPs) with arbitrary phase space of states [14–16].

Definition [16]. Semi-Markov kernel (SM-kernel) in a measurable space (E, $\mathcal{B}$) is the function $Q(t,x,B)$, satisfying the conditions:

(1) $Q(t,x,B)$ are nondecreasing right-continuous functions of $t \geq 0, Q(0,x,B)=0, x \in E,\ B \in \mathcal{B}$;

(2) with $t>0$ fixed, $Q(t,x,B)$ is a semistochastic kernel: $Q(t,x,B) \leq 1$;

(3) $Q(+\infty,x,B)$ is a stochastic kernel by x, B, that is $Q(+\infty,x,E) \equiv 1$.

An SMP with arbitrary phase space of states is defined by means of a Markov renewal process (MRP).

Definition [16]. An MRP is a two-dimensional Markov chain $\{\xi_n,\theta_n;n \geq 0\}$ taking values in $E \times (0,\infty)$. Its transition probabilities are given by the expression:

$$P\{\xi_{n+1} \in B, \theta_{n+1} \leq t / \xi_n = x\} = Q(t,x,B),$$

where $Q(t,x,B)$ is an SM-kernel in (E, $\mathcal{B}$).

The first component $\{\xi_n;n \geq 0\}$ of the MRP $\{\xi_n,\theta_n;n \geq 0\}$ is a Markov chain. Its transition probabilities are defined by means of SM-kernel $Q(t,x,B)$:

$$P(x,B) = P\{\xi_{n+1} \in B / \xi_n = x\} = Q(+\infty,x,B).$$

It is called an embedded Markov chain (EMC) of MRP $\{\xi_n,\theta_n;n \geq 0\}$. RVs $\theta_n, n \geq 0$, making the second component of MRP $\{\xi_n,\theta_n;n \geq 0\}$, determine intervals between the moments τ_n of Markov restoration:

$$\tau_n = \sum_{k=1}^{n} \theta_k\ n \geq 1, \tau_0 = 0.$$

Consider the counting process $\nu(t): \nu(t) = \sup\{n : \tau_n \leq t\}$, which counts the number of Markov renewal moments in $[0,t]$.

Definition [16]. The process $\xi(t) = \xi_{\nu(t)}$ is an SMP corresponding MRP $\{\xi_n,\theta_n;n \geq 0\}$.

It can be concluded from the definition that SMP is a jump right-continuous process: $\xi(t+0) = \xi(t)$.

Another way of SMP definition is the following [16]:

(1) stochastic kernel

$$P(x,B) = P\{\xi_{n+1} \in B / \xi_n = x\}, x \in E, B \in \mathcal{B}$$

(2) DF of sojourn times of EMC $\{\xi_n;n \geq 0\}$ transitions

$$G(t,x,y) = G_{xy}(t) = P\{\theta_{n+1} \leq t / \xi_n = x, \xi_{n+1} = y\}$$

are defined.

Then SM-kernel $Q(t,x,B)$ is defined by the formula [16]:

$$Q(t,x,B)=\int_B G(t,x,y)P(x,dy). \tag{1.14}$$

Let us write out definitions and formulas of some reliability and efficiency characteristics of restorable systems described by means of SMP.

Let a system S be described by SMP $\xi(t)$ with a phase space $(E, \mathcal{B})$. Assume the set of SMP $\xi(t)$ states can be represented as

$$E=E_+\bigcup E_-, E_+\bigcap E_-=\varnothing, E_+\in B, E_-\in \mathcal{B},$$

where E_+ and E_- are interpreted as sets of system S up- and down-states, respectively.

Definition [16]. Stationary availability factor K_a of system S is the number, given by

$$K_a=\lim_{t\to+\infty} P\{\xi(t)\in E_+ / \xi(0)=x\},$$

under assumption the limit existence and independence on the initial state $x\in E$.

The following stationary reliability characteristics of restorable systems are often in use. Their formal definition is given in [16]:

(a) mean stationary operating time to failure T_+,
(b) mean stationary restoration time T_-.

EMC $\{\xi_n;n\ge 0\}$ stationary distribution $\rho(B)$ satisfies the integral equation:

$$\rho(B)=\int_E \rho(dx)P(x,B), B\in \mathcal{B}. \tag{1.15}$$

It was proved in [16], that if the unique stationary distribution of EMC $\{\xi_n;n\ge 0\}$ of SMP $\xi(t)$ describing the system S operation exists, characteristics K_a,T_+,T_- are given by the formulas:

$$K_a=\frac{\int_{E_+} m(x)\rho(dx)}{\int_E m(x)\rho(dx)}, \tag{1.16}$$

$$T_+=\frac{\int_{E_+} m(x)\rho(dx)}{\int_{E_+} P(x,E_-)\rho(dx)}, \tag{1.17}$$

$$T_- = \frac{\int_{E_-} m(x)\rho(dx)}{\int_{E_+} P(x,E_-)\rho(dx)}, \tag{1.18}$$

under some assumptions.

Here $\rho(dx)$ denotes the EMC $\{\xi_n; n \geq 0\}$ stationary distribution, and $m(x)$ is the mean sojourn time in state $x \in E$. One should note the characteristics K_a, T_+, T_- relate like this:

$$K_a = \frac{T_+}{T_+ + T_-}. \tag{1.19}$$

The Markov renewal equation [16] plays an important role in the theory of SMP. It is as follows:

$$u(x,t) = g(x,t) + \int_0^t \int_E Q(ds,x,dy)u(y,t-s), x \in E, t \geq 0. \tag{1.20}$$

Markov renewal equations for some SMP characteristics are given in [16]. The Markov renewal equation for the distribution of the sojourn time $\overline{R}_x(t)$ of SMP $\xi(t)$ in a certain subset E_0 of states is often applied [16]:

$$\overline{R_x}(t) = \overline{F_x}(t) + \int_0^t \int_{E_0} Q(ds,x,dy)\overline{R_y}(t-s), \tag{1.21}$$

its consequence is the equation for mean sojourn times in a subset E_0 [15]:

$$U(x) = \int_{E_0} P(x,dy)U(y) + m(x),$$

where $m(x)$ is the SMP $\xi(t)$ mean sojourn time in x.

Stationary efficiency characteristics of system operation are: S is the mean specific income per calendar time unit and C is the mean specific expenses per time unit of up-state. In terms of SM model, these characteristics are given by the ratios [18,26]:

$$S = \frac{\int_E m(x)f_s(x)\rho(dx)}{\int_E m(x)\rho(dx)}, \tag{1.22}$$

$$C = \frac{\int_E m(x)f_c(x)\rho(dx)}{\int_{E_+} m(x)\rho(dx)}, \tag{1.23}$$

where $f_s(x)$, $f_c(x)$ are functions denoting income and expenses in each state.

In the monograph, the following method of approximation of system stationary reliability characteristics, introduced in [14], is applied.

Let the initial system S operation is described by SMP $\xi(t)$ with a phase space $(E, \mathcal{B})$. The set E of states is divided into two subsets E_+ and E_-, so that $E = E_+ \cup E_-, E_+ \cap E_- = \varnothing$. Assume the kernel $P(x,B)$, $B \in \mathcal{B}$, of EMC $\{\xi_n; n \geq 0\}$ of SMP $\xi(t)$ is close to the kernel $P^{(0)}(x,B)$, $B \in \mathcal{B}$, of EMC $\{\xi_n^{(0)}; n \geq 0\}$ of supporting system $S^{(0)}$ having unique stationary distribution $\rho^{(0)}(B)$, $B \in, \mathcal{B}$.

Then instead of the expressions (1.17) and (1.18) we can use the following formulas [14]:

$$T_+ \approx \frac{\int_{E_+} m(x)\rho^{(0)}(dx)}{\int_{E_+} P^{(r)}(x,E_-)\rho^{(0)}(dx)}, T_- \approx \frac{\int_E \rho^{(0)}(dx) \int_{E_-} m(y) P^{(r)}(x,dy)}{\int_{E_+} P^{(r)}(x,E_-)\rho^{(0)}(dx)}, \tag{1.24}$$

approximating characteristics of the initial system S.

Here, $\rho^{(0)}(dx)$ is the EMC $\{\xi_n^{(0)}; n \geq 0\}$ stationary distribution for supporting system; $m(x)$ is the mean sojourn times in the states of the initial system; $P^{(r)}(x,E_-)$ is the the probabilities of EMC $\{\xi_n; n \geq 0\}$ transitions from up- into down-states in minimal path for the initial system; r is a minimum of steps, necessary for transition from the states of E_+, belonging to the ergodic class $E^{(0)}$ of the initial system, to the set of down-states E_-. Under $r = 1$, formula (1.24) takes the form:

$$T_+ \approx \frac{\int_{E_+} m(x)\rho^{(0)}(dx)}{\int_{E_+} P(x,E_-)\rho^{(0)}(dx)}, T_- \approx \frac{\int_E \rho^{(0)}(dx) \int_{E_-} m(y) P(x,dy)}{\int_{E_+} P(x,E_-)\rho^{(0)}(dx)}. \tag{1.25}$$

The kernel $P(x,B)$ of the initial system EMC $\{\xi_n; n \geq 0\}$ is close to the kernel $P^{(0)}(x,B)$ of supporting system EMC $\{\xi_n; n \geq 0\}$, that is why under $r = 1$, along with the second formula (1.25), the following approximating formula for T_- can be used:

$$T_- \approx \frac{\int_{E_-} m(x)\rho^{(0)}(dx)}{\int_{E_+} P(x,E_-)\rho^{(0)}(dx)}. \tag{1.26}$$

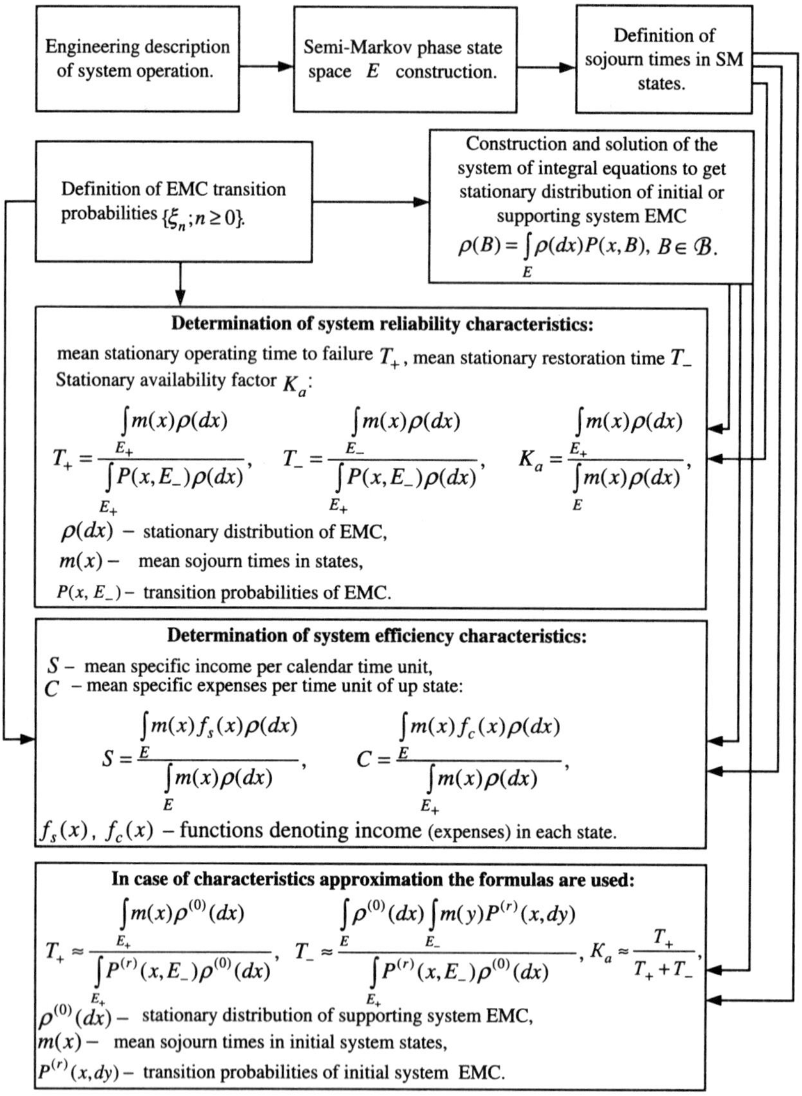

FIGURE 1.2 General scheme of semi-Markov model building

To approximate system stationary efficiency characteristics, instead of (1.22) and (1.23) the following ratios will be used:

$$S \approx \frac{\int_E m(x) f_s(x) \rho^{(0)}(dx)}{\int_E m(x) \rho^{(0)}(dx)}, C \approx \frac{\int_E m(x) f_c(x) \rho^{(0)}(dx)}{\int_{E_+} m(x) \rho^{(0)}(dx)}, \tag{1.27}$$

where $\rho^{(0)}(dx)$ is the stationary distribution of supporting system EMC $\{\xi_n^{(0)}; n \geq 0\}$; $m(x)$ is the mean sojourn times in the states of the initial system; and $f_s(x)$, $f_c(x)$ are the functions denoting income and expenses in each state of the initial system.

Semi-Markov models of latent failures control are built under the following assumptions:

(1) From the point of view of reliability, a system component is a minimal compound element (detail), which can be failed, controlled, and restored.
(2) Component failure is detected while control execution only.
(3) After failure detection, restoration process immediately begins.
(4) A component is completely restored while restoration process.
(5) DFs of the RVs: operating time to failure, time periods between the moments of control execution, control and restoration time are arbitrary ones.

The stages of semi-Markov model construction and system stationary characteristics definition are represented in Figure 1.2.

Chapter 2

Semi-Markov Models of One-Component Systems with Regard to Control of Latent Failures

Chapter Outline

Semi-Markov Models. http://dx.doi.org/10.1016/B978-0-12-802212-2.00002-4
Copyright © 2015 Elsevier Inc. All rights reserved.

2.1 THE SYSTEM MODEL WITH COMPONENT DEACTIVATION WHILE CONTROL EXECUTION

2.1.1 The System Description

The system S consists of one functional component and control unit. The system operates as follows. At the time zero, the component begins operating, and control is activated. Component time to failure (TF) is random variable (RV) α with distribution function (DF) $F(t) = P\{\alpha \le t\}$ and distribution density (DD) $f(t)$. Control is executed in random time δ with DF $R(t) = P\{\delta \le t\}$ and DD $r(t)$. Failure is detected after control execution only (latent failure). While control execution, component does not operate. Control execution time is RV γ with DF $V(t) = P\{\gamma \le t\}$ and DD $v(t)$. Component restoration time (RT) is RV β with DF $G(t) = P\{\beta \le t\}$ and DD $g(t)$. Control is suspended for the restoration period. After restoration, all the component characteristics are renewed. RV α, β, δ, γ are assumed to be independent and to have finite expectations.

2.1.2 Semi-Markov Model Building

Let us describe the system operation by means of semi-Markov process $\xi(t)$ with discrete–continuous phase state space. Introduce the following set E of semi-Markov system states:

$$E = \{111, 2\hat{1}0x, 211x, 101x, 2\hat{0}0, 222\}.$$

The meaning of state codes is the following:

111 – component begins operating, control is activated;
$2\hat{1}0x$ – control has begun, the component is in up-state, has been deactivated, time $x > 0$ is left till failure (regardless of control execution time);
$211x$ – control has ended, the component continues operating, time $x > 0$ is left till failure;
$101x$ – failure has occurred, time $x > 0$ is left till the beginning of control;
$2\hat{0}0$ – control has begun, failed component has been deactivated;
222 – control has ended, failure is detected, component restoration has begun.

Time diagram and system transition graph are given in Figures 2.1 and 2.2, respectively. Latent failures are marked with a broken line, whereas component deactivation and control time suspension are in bold.

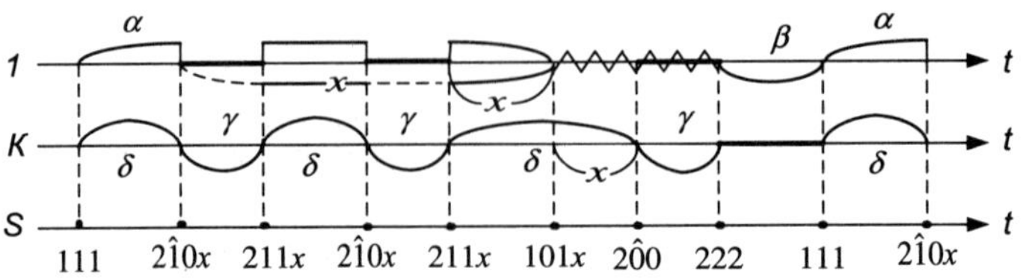

FIGURE 2.1 Time diagram of system operation

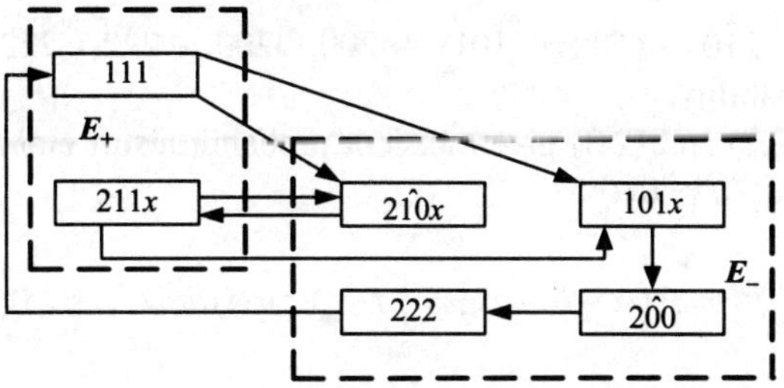

FIGURE 2.2 System transition graph

Let us define system sojourn times in states. For instance, system sojourn time θ_{211x} in $211x$ is determined by two factors: residual time x till the latent failure and time δ of control periodicity.

Consequently, $\theta_{211x} = \delta \wedge x$, where $\wedge$ is a minimum sign. Sojourn times in other states are defined in the same way:

$$\theta_{111} = \alpha \wedge \delta, \theta_{2\hat{1}0x} = \gamma, \theta_{101x} = x, \theta_{2\hat{0}0} = \gamma, \theta_{222} = \beta. \tag{2.1}$$

Let us describe system transition events. Transition events from the states 111, $211x$ are illustrated in Figures 2.3 and 2.4, respectively. They are defined by expressions (2.2) and (2.3), respectively.

$$\begin{aligned}&\{111 \to 2\hat{1}0dx\} = \{\alpha - \delta \in dx\}, x > 0;\\&\{111 \to 101dx\} = \{\delta - \alpha \in dx\}, x > 0.\end{aligned} \tag{2.2}$$

$$\begin{aligned}&\{211x \to 2\hat{1}0dy\} = \{x - \delta \in dy\}, 0 < y < x;\\&\{211x \to 101dy\} = \{\delta - x \in dy\}, y > 0.\end{aligned} \tag{2.3}$$

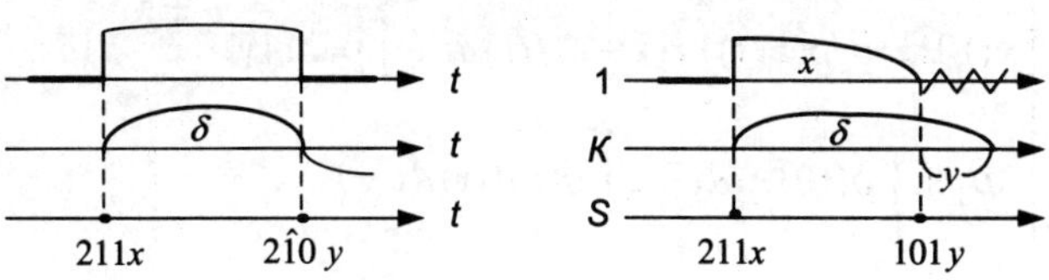

FIGURE 2.3 Transition events from the state 111

FIGURE 2.4 Transition events from the state $211x$

Transitions $2\hat{1}0x \to 211x$, $101x \to 2\hat{0}0$, $2\hat{0}0 \to 222$, $222 \to 111$ occur with unity probability.

Formulas (2.2) and (2.3) give transient probabilities of embedded Markov chain (EMC) $\{\xi_n;\, n \geq 0\}$:

$$\begin{aligned}
&p_{111}^{2\hat{1}0dx} = P\{\alpha - \delta \in dx\} = \int_0^\infty f(x+t)r(t)\,dt\,dx, \quad x>0;\\
&p_{111}^{101dx} = P\{\delta - \alpha \in dx\} = \int_0^\infty r(x+t)f(t)\,dt\,dx, \quad x>0;\\
&p_{211x}^{2\hat{1}0dy} = P\{x - \delta \in dy\} = r(x-y)dy, \quad 0<y<x;\\
&p_{211x}^{101dx} = P\{\delta - x \in dy\} = r(x+y)dy, \quad y>0;\\
&P_{2\hat{1}0x}^{211x} = P_{101x}^{2\hat{0}0} = P_{2\hat{0}0}^{222} = P_{222}^{111} = 1.
\end{aligned} \tag{2.4}$$

It is no use defining semi-Markov kernel $Q(t,x,\mathrm{B})$ of Markov renewal process $\{\xi_n, \theta_n; n \geq 0\}$ by the formula (1.14) to obtain system stationary characteristics.

2.1.3 Definition of EMC Stationary Distribution

Let us denote by $\rho(111)$, $\rho(2\hat{0}0)$, $\rho(222)$ the values of EMC $\{\xi_n;\, n \geq 0\}$ stationary distribution in states 111, $2\hat{0}0$, 222. Assume stationary densities $\rho(2\hat{1}0x)$, $\rho(211x)$, and $\rho(101x)$ exist for states $2\hat{1}0x$, $211x$, and $101x$, respectively.

Using probabilities and probability densities of EMC $\{\xi_n;\, n \geq 0\}$ transition (2.4), taking into account (1.15), let us construct the system of integral equations for the stationary distribution definition.

$$\begin{cases}
\rho_0 = \rho(111) = \rho(222),\\
\rho(222) = \rho(2\hat{0}0),\\
\rho(2\hat{0}0) = \int_0^\infty \rho(101x)\,dx,\\
\rho(2\hat{1}0x) = \rho(111)\int_0^\infty f(x+t)r(t)\,dt + \int_x^\infty \rho(211y)r(y-x)\,dy,\\
\rho(211x) = \rho(2\hat{1}0x),\\
\rho(101x) = \rho(111)\int_0^\infty r(x+t)f(t)\,dt + \int_0^\infty \rho(211y)r(y+x)\,dy,\\
3\rho_0 + \int_0^\infty \rho(101x)\,dx + 2\int_0^\infty \rho(2\hat{1}0x)\,dx = 1.
\end{cases} \tag{2.5}$$

The last equation in the system (2.5) is a normalization requirement.

By means of the method of successive approximations [11], the system of equations (2.5) is proved to have the following solutions:

$$\begin{cases} \rho_0 = \rho(111) = \rho(222) = \rho(2\hat{0}0), \\ \rho(211x) = \rho(2\hat{1}0x) = \rho_0 \int\limits_0^\infty h_r(t) f(x+t)dt, \\ \rho(101x) = \rho_0 \int\limits_0^\infty v_r(z,x) f(z)\,dz, \end{cases} \tag{2.6}$$

where, according to (1.2) and (1.8), $h_r(t) = \sum\limits_{n=1}^{\infty} r^{*(n)}(t)$ is the density of renewal function $H_r(t)$ of renewal process, generated by RV δ; $r^{*(n)}(t)$ is the nth fold convolution of the DD $r(t)$; and $v_r(z,x) = r(z+x) + \int\limits_0^z r(z+x-s)h_r(s)ds$ is the DD of the direct residual time for the same renewal process. The value of constant ρ_0 is obtained from the normalization requirement.

2.1.4 Stationary Characteristics Definition

Let us split phase state space E into the following two subsets:

$E_+ = \{111, 211x\}$ – the system is in up-state;

$E_- = \{2\hat{1}0x, 101x, 2\hat{0}0, 222\}$ – the system is in down-state.

Let us find average stationary operating TF T_+ and average stationary RT T_- with the help of formulas (1.17) and (1.18), respectively:

$$T_+ = \frac{\int\limits_{E_+} m(e)\rho(de)}{\int\limits_{E_+} P(e,E_-)\rho(de)}, \quad T_- = \frac{\int\limits_{E_-} m(e)\rho(de)}{\int\limits_{E_+} P(e,E_-)\rho(de)}, \tag{2.7}$$

where $\rho(de)$ is the EMC $\{\xi_n; n \geq 0\}$ stationary distribution, $m(e)$ are average values of system sojourn times in states, and $P(e,E_-)$ are the probability transitions of EMC $\{\xi_n; n \geq 0\}$ from up- into down-states.

Let us determine system average sojourn times in states with the help of formula (2.1):

$$\begin{aligned} &m(111) = \int\limits_0^\infty \bar{F}(t)\bar{R}(t)\,dt, \bar{F}(t) = 1 - F(t), \bar{R}(t) = 1 - R(t), m(211x) = \int\limits_0^x \bar{R}(t)\,dt, \\ &m(2\hat{1}0x) = E\gamma, m(101x) = x, m(2\hat{0}0) = E\gamma, m(222) = E\beta. \end{aligned} \tag{2.8}$$

Taking into account (2.4), (2.6), and (2.8), let us write down the expressions included in (2.7).

$$\int_{E_+} m(e)\rho(de) = m(111)\rho(111) + \int_0^\infty m(211x)\rho(211x)dx =$$
$$= \rho_0\left(\int_0^\infty \overline{F}(t)\overline{R}(t)dt + \int_0^\infty dx \int_0^x \overline{R}(t)dt \int_0^\infty h_r(y)f(x+y)dy\right) = \tag{2.9}$$
$$= \rho_0\left(\int_0^\infty \overline{F}(t)\overline{R}(t)dt + \int_0^\infty \overline{R}(t)dt \int_t^\infty dx \int_0^\infty h_r(y)f(x+y)dy\right).$$

Changing the order of integration and integrating in parts the second summand of the expression (2.9), we get:

$$\int_{E_+} m(e)\rho(de) = \rho_0 E\alpha.$$

Next,

$$\int_{E_-} m(e)\rho(de) = m(2\hat{0}0)\rho(2\hat{0}0) + m(222)\rho(222) +$$
$$+\int_0^\infty m(2\hat{1}0x)\rho(2\hat{1}0x)dx + \int_0^\infty m(101x)\rho(101x)dx =$$
$$= \rho_0\left(E\gamma + E\beta + E\gamma\int_0^\infty dx\int_0^\infty h_r(t)f(x+t)dt + \int_0^\infty xdx\int_0^\infty v_r(z,x)f(z)dz\right) =$$
$$= \rho_0\left(E\gamma + E\beta + E\gamma\int_0^\infty h_r(t)dt\int_0^\infty f(x+t)dx + \int_0^\infty f(z)dz\int_0^\infty xv_r(z,x)dx\right) =$$
$$= \rho_0\left(E\gamma + E\beta + E\gamma\int_0^\infty \overline{F}(t)h_r(t)dt + \int_0^\infty f(z)\left(E\delta\tilde{H}_r(z) - z\right)dz\right) =$$
$$= \rho_0\left(E\gamma + E\beta - E\alpha + E\gamma\int_0^\infty \overline{F}(t)h_r(t)dt + E\delta\int_0^\infty \tilde{H}_r(z)f(z)dz\right).$$

Integrating in parts the fourth summand of the latter expression, we have:

$$\int_{E_-} m(e)\rho(de) = \rho_0\left(E\beta - E\alpha + \left(E\delta + E\gamma\right)\int_0^\infty \tilde{H}_r(t)f(t)dt\right),$$

where, according to (1.1) and (1.4), $\tilde{H}_r(t) = 1 + H_r(t)$, $H_r(t) = \sum_{n=1}^{\infty} R^{*(n)}(t)$ is the renewal function of the renewal process, generated by RV δ, and $R^{*(n)}(t)$ is the nth fold convolution of the DF $R(t)$.

Then,

$$\int_E m(e)\rho(de) = \rho_0\left(E\beta + (E\delta + E\gamma)\int_0^\infty \tilde{H}_r(t)f(t)\,dt\right).$$

Next,

$$\int_{E_+} P(e,E_-)\rho(de) = P(111,E_-)\rho(111) + \int_0^\infty P(211x,E_-)\rho(211x)\,dx =$$

$$= \rho_0\left(\int_0^\infty dx\int_0^\infty f(x+t)r(t)\,dt + \int_0^\infty dx\int_0^\infty r(x+t)f(t)\,dt +\right.$$

$$\left.+\int_0^\infty dx\int_0^\infty h_r(z)f(x+z)\,dz\int_0^x r(x-y)\,dy + \int_0^\infty dx\int_0^\infty h_r(z)f(x+z)\,dz\int_0^\infty r(x+y)\,dy\right) =$$

$$= \rho_0\left(1 + \int_0^\infty R(x)\,dx\int_0^\infty h_r(z)f(x+z)\,dz + \int_0^\infty \overline{R}(x)\,dx\int_0^\infty h_r(z)f(x+z)\,dz\right) =$$

$$= \rho_0\left(1 + \int_0^\infty h_r(z)\,dz\int_0^\infty f(x+z)\,dx\right) = \rho_0\left(1 + \int_0^\infty \overline{F}(z)h_r(z)dz\right) = \rho_0\int_0^\infty \tilde{H}_r(z)f(z)\,dz.$$

Therefore, average stationary operating TF T_+ is

$$T_+ = \frac{E\alpha}{\int_0^\infty \tilde{H}_r(t)f(t)\,dt}. \tag{2.10}$$

Average stationary RT T_- is

$$T_- = \frac{E\beta - E\alpha + (E\delta + E\gamma)\int_0^\infty \tilde{H}_r(t)f(t)\,dt}{\int_0^\infty \tilde{H}_r(t)f(t)\,dt}. \tag{2.11}$$

Let us find stationary availability factor from the ratio (1.19).

$$K_a = \frac{E\alpha}{E\beta + (E\delta + E\gamma)\int_0^\infty \tilde{H}_r(t)f(t)\,dt}. \tag{2.12}$$

In terms of the function $\hat{H}_r(t) = \begin{cases} 1 + H_r(t), t > 0, \\ 0, t = 0, \end{cases}$ stationary availability factor can be given as follows:

$$K_a = \frac{E\alpha}{E\beta + (E\delta + E\gamma)\int_0^\infty \overline{F}(t)\,d\hat{H}_r(t)}, \tag{2.13}$$

where the integration is carried out over the interval $[0; \ +\infty)$.

In the same way other characteristics can be obtained.

Let us define system stationary efficiency characteristics: average specific income per calendar time unit S and average specific expenses per time unit of up-state C [14]. The formulas (1.22) and (1.23) are applied:

$$S=\frac{\int_E m(e)f_s(e)\rho(de)}{\int_E m(e)\rho(de)},C=\frac{\int_E m(e)f_c(e)\rho(de)}{\int_{E_+} m(e)\rho(de)}.$$

Here $f_s(e), f_c(e)$ are the functions, defining income and expenses per time unit in each state, respectively.

For the given system, functions $f_s(e), f_c(e)$ are as follows:

$$f_s(e)=\begin{cases} c_1, & e\in\{111,211x\},\\ -c_2, & e=222,\\ -c_3, & e\in\{2\hat{1}0x,2\hat{0}0\},\\ -c_4, & e=101x, \end{cases} \qquad f_c(e)=\begin{cases} 0, & e\in\{111,211x\},\\ c_2, & e=222,\\ c_3, & e\in\{2\hat{1}0x,2\hat{0}0\},\\ c_4, & e=101x. \end{cases} \tag{2.14}$$

Here c_1 is the income per time unit of the component up-state, c_2 denotes expenses per time unit of restoration, c_3 are expenses per time unit of control, and c_4 are wastes per time unit of latent failure of the component.

Taking into account formulas (2.4), (2.6), (2.8), and (2.14), average specific income per calendar time unit is defined by the ratio:

$$S=\frac{E\alpha(c_1+c_4)-c_2E\beta-(c_3E\gamma+c_4E\delta)\int_0^\infty \tilde{H}_r(t)f(t)\,dt}{E\beta+(E\delta+E\gamma)\int_0^\infty \tilde{H}_r(t)f(t)\,dt}. \tag{2.15}$$

Average specific expenses per time unit of system up-state are as follows:

$$C=\frac{c_2E\beta-c_4E\alpha+(c_4E\delta+c_3E\gamma)\int_0^\infty \tilde{H}_r(t)f(t)\,dt}{E\alpha}. \tag{2.16}$$

Let us get formulas for reliability and efficiency characteristics, under the condition that time periods between control are nonrandom values $\tau>0$,

$\tau = \text{const}$. Taking into account that in this case $R(t) = 1(t - \tau)$, $E\delta = T$, expressions (2.10)–(2.12), (2.15), and (2.16) transform into

$$T_+ = \frac{E\alpha}{\sum_{n=0}^{\infty} \overline{F}(n\tau)}, \quad T_- = \frac{E\beta - E\alpha + (\tau + E\gamma)\sum_{n=0}^{\infty} \overline{F}(n\tau)}{\sum_{n=0}^{\infty} \overline{F}(n\tau)},$$
$$K_a = \frac{E\alpha}{E\beta + (\tau + E\gamma)\sum_{n=0}^{\infty} \overline{F}(n\tau)}, \tag{2.17}$$

$$S = \frac{E\alpha(c_1 + c_4) - c_2 E\beta - (c_3 E\gamma + c_4\tau)\sum_{n=0}^{\infty} \overline{F}(n\tau)}{E\beta + (\tau + E\gamma)\sum_{n=0}^{\infty} \overline{F}(n\tau)}, \tag{2.18}$$

$$C = \frac{c_2 E\beta - c_4 E\alpha + (c_4\tau + c_3 E\gamma)\sum_{n=0}^{\infty} \overline{F}(n\tau)}{E\alpha}. \tag{2.19}$$

The above formulas allow to calculate the stationary steady-state availability factor, average income and expenses for different initial data and to solve problems of control execution periodicity optimization.

We should note, under $E\gamma \to 0$ and $\tau \to 0$, we get characteristics for the system with continuous control:

$$K_a = \frac{E\alpha}{E\alpha + E\beta}, \quad S = \frac{c_1 E\alpha - c_2 E\beta}{E\alpha + E\beta}, \quad C = c_2 \frac{E\beta}{E\alpha}.$$

Example. Initial data are as follows: control periodicity is $\tau = 5\text{h}$, average TF $E\alpha$, h, average RT $E\beta$, h, average control duration $E\gamma$, h; the distributions of RV α are as follows: exponential, Erlangian of the 4th and 8th order, Veibull–Gnedenko with shape parameter $\beta = 2$. Here $c_1 = 7$ c.u., $c_2 = 2$ c.u., $c_3 = 3$ c.u., $c_4 = 1$ c.u.

Initial data and calculation results, obtained by (2.17), (2.18), and (2.19) for the given system, are represented in Table 2.1.

TABLE 2.1 Values of $K_a(\tau)$, $S(\tau)$, and $C(\tau)$, Under $\tau = 5$ h

Initial data				Results		
Kind of distribution of RV α	$E\alpha$, h	$E\beta$, h	$E\gamma$, h	$K_a(\tau)$	$S(\tau)$, c.u./h	$C(\tau)$, c.u./h
Exponential	60	0.083	0.10	0.939	6.435	0.150
Exponential	70	0.500	0.30	0.898	6.121	0.192
Erlangian of the 4th order	60	0.083	0.10	0.939	6.400	0.140
Erlangian of the 8th order	70	0.500	0.30	0.899	6.124	0.190
Veibull–Gnedenko	60	0.083	0.10	0.916	6.440	0.149
Veibull–Gnedenko	70	0.500	0.30	0.905	6.121	0.236

2.2 THE SYSTEM MODEL WITHOUT COMPONENT DEACTIVATION WHILE CONTROL EXECUTION

2.2.1 The System Description

The given system S consists of one component performing definite functions and of equipment, which controls its operability. Let us describe its operation. At the time zero component begins operating, and control is activated. Component TF is RV α with DF $F(t) = P\{\alpha \le t\}$ and DD $f(t)$. Control is executed in random time δ with DF $R(t) = P\{\delta \le t\}$ and DD $r(t)$. Component failure is detected while control (latent failure). Component operation is not suspended for the control period. Duration of control execution is RV γ with DF $V(t) = P\{\gamma \le t\}$ and DD $v(t)$. Component restoration begins after failure detection, and control is deactivated. Component RT is RV β with DF $G(t) = P\{\beta \le t\}$ and DD $g(t)$. After restoration, all the component characteristics are renewed completely. The operation begins, and control is activated. RV α, β, δ, and γ are assumed to be independent and have finite expectations.

2.2.2 Semi-Markov Model Building

Let us describe the system operation by means of semi-Markov process $\xi(t)$ with discrete–continuous phase state space. Introduce the following set E of semi-Markov system states:

$$E = \{111, 210x, 211x, 100x, 222, 101x, 200\}.$$

The essence of state codes is as follows:

111 – component begins operating, control is activated;
210x – control has begun, component is in up-state, time $x > 0$ is left till the latent failure;
211x – control has ended, component is in up-state, time $x > 0$ is left till the latent failure;

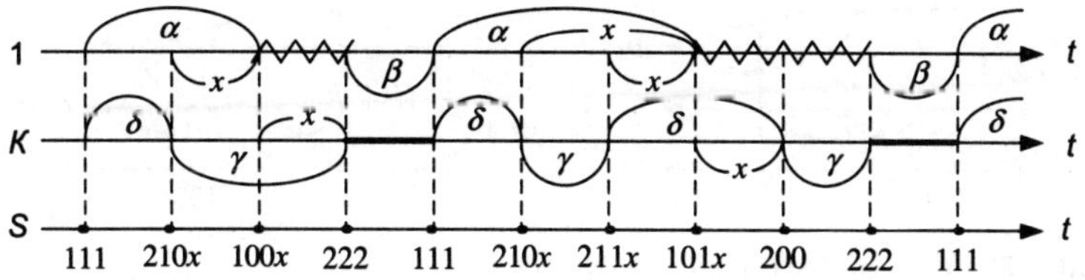

FIGURE 2.5 Time diagram of system operation

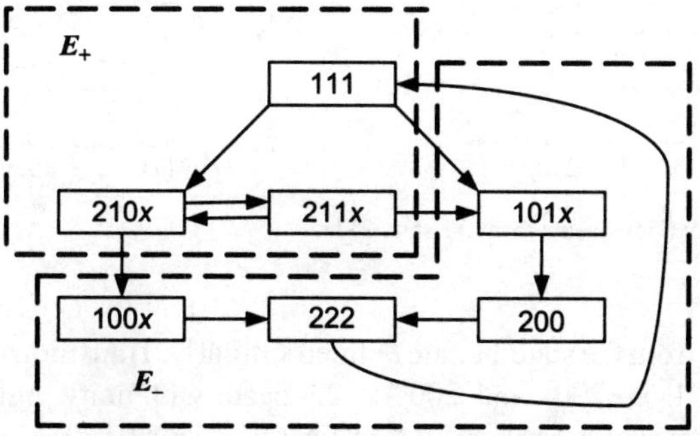

FIGURE 2.6 System transition graph

$100x$ – component failure has occurred, control is carried out, time $x > 0$ is left till the latent failure detection;
222 – control has been suspended, failure has been detected, its restoration begins;
$101x$ – component has failed, time $x > 0$ is left till the beginning of control;
200 – control has begun, component is in down-state.

Time diagram of system operation and system transition graph are shown in Figures 2.5 and 2.6, respectively.

Let us find system sojourn times in its states. For example, system sojourn time θ_{211x} in $211x$ is determined by two factors: time x is left till the latent failure and time γ defines control duration. Consequently, $\theta_{211x} = \gamma \wedge x$. Sojourn times in other states are defined similarly:

$$\theta_{111} = \alpha \wedge \delta, \theta_{211x} = \delta \wedge x, \theta_{100x} = x, \theta_{222} = \beta, \theta_{101x} = x, \theta_{200} = \gamma. \quad (2.20)$$

Let us describe state transition events. Thus, transition events from $210x$ and $211x$ are shown in Figures 2.7 and 2.8, respectively. They are defined by ratios (2.21) and (2.22), respectively.

$$\begin{aligned} \{210x \to 211dy\} &= \{x - \gamma \in dy\}, 0 < y < x; \\ \{210x \to 100dy\} &= \{\gamma - x \in dy\}, y > 0. \end{aligned} \quad (2.21)$$

$$\begin{aligned} \{211x \to 210dy\} &= \{x - \delta \in dy\}, 0 < y < x; \\ \{211x \to 101dy\} &= \{\delta - x \in dy\}, y > 0. \end{aligned} \quad (2.22)$$

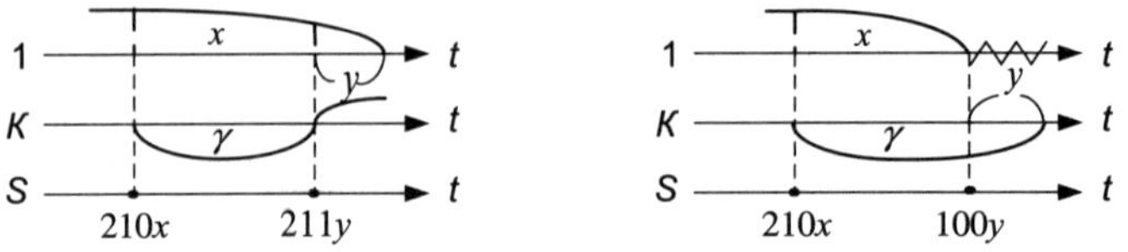

FIGURE 2.7 Transition events from the state $210x$

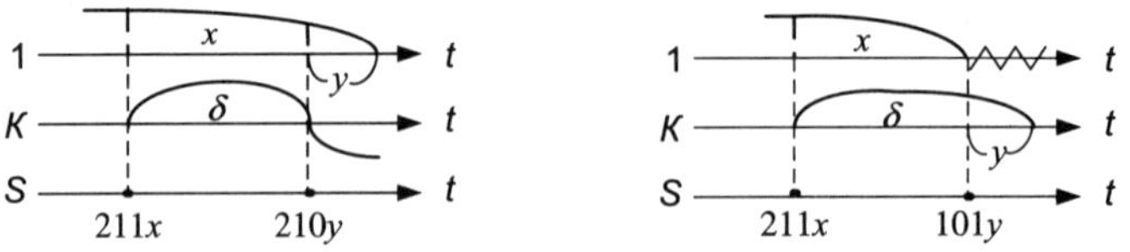

FIGURE 2.8 Transition events from the state $211x$

Transitions from the state 111 are defined similarly. Transitions $100x \to 222$, $222 \to 111$, $101x \to 200$, and $200 \to 222$ occur with unity probability.

Let us find transition probabilities of EMC $\{\xi_n; n \geq 0\}$. Taking into account formulas (2.21) and (2.22) we get the following:

$$\begin{aligned}
p_{111}^{210dx} &= P\{\alpha - \delta \in dx\} = \int_0^\infty f(x+t)r(t)\,dt\,dx, \quad x > 0; \\
p_{111}^{101dx} &= P\{\delta - \alpha \in dx\} = \int_0^\infty r(x+t)f(t)\,dt\,dx, \quad x > 0; \\
p_{210x}^{100dy} &= P\{\gamma - x \in dy\} = v(y+x)dy, \quad y > 0; \\
p_{211x}^{210dy} &= P\{x - \gamma \in dy\} = v(x-y)dy, \quad 0 < y < x; \\
p_{210x}^{211dy} &= P\{\delta - x \in dy\} = r(x-y)dy, \quad 0 < y < x; \\
p_{211x}^{101dy} &= P\{x - \delta \in dy\} = r(y+x)dy, \quad y > 0; \\
P_{100x}^{222} &= P_{222}^{111} = P_{101x}^{200} = P_{200}^{222} = 1.
\end{aligned} \tag{2.23}$$

2.2.3 Definition of EMC Stationary Distribution

Let us denote by $\rho(111)$, $\rho(222)$, and $\rho(200)$ the values of EMC stationary distribution $\{\xi_n; n \geq 0\}$ in states 111, 222, and 200, respectively, and assume existence of stationary densities $\rho(210x)$, $\rho(211x)$, $\rho(100x)$, and $\rho(101x)$ for states $210x$, $211x$, $100x$, and $101x$, respectively.

Using probabilities and densities of transition probabilities of EMC $\{\xi_n; n \geq 0\}$ (2.23), taking into account formula (1.15), let us construct the system of integral equations for the stationary distribution definition:

$$
\begin{cases}
\rho_0 = \rho(111) = \rho(222), \\
\rho(210x) = \rho(111)\int\limits_0^\infty f(x+t)r(t)\,dt + \int\limits_x^\infty \rho(211y)r(y-x)\,dy, \\
\rho(211x) = \int\limits_x^\infty \rho(210y)v(y-x)\,dy, \\
\rho(100x) = \int\limits_0^\infty \rho(210y)v(x+y)\,dy, \\
\rho(222) = \int\limits_0^\infty \rho(100x)\,dx + \rho(200), \\
\rho(101x) = \rho(111)\int\limits_0^\infty r(x+t)f(t)\,dt + \int\limits_0^\infty \rho(211y)r(y+x)\,dy, \\
\rho(200) = \int\limits_0^\infty \rho(101x)\,dx, \\
2\rho_0 + \rho(200) + \int\limits_0^\infty \left(\rho(210x) + \rho(100x) + \rho(211x) + \rho(101x)\right) dx = 1.
\end{cases}
\tag{2.24}
$$

The last equation of the system (2.24) is a normalization requirement.

By means of the method of successive approximations [9], the system (2.24) is shown to have the following solutions (Appendix A):

$$
\begin{cases}
\rho(111) = \rho(222) = \rho_0, \\
\rho(210x) = \rho_0 \int\limits_x^\infty h^{(0)}(t-x)f(t)\,dt, \\
\rho(211x) = \rho_0 \int\limits_x^\infty h^{(1)}(t-x)f(t)\,dt, \\
\rho(100x) = \rho_0 \int\limits_0^\infty f(t)v^{(0)}(t,x)\,dt, \\
\rho(101x) = \rho_0 \int\limits_0^\infty f(t)v^{(1)}(t,x)\,dt, \\
\rho(200) = \rho_0 \int\limits_0^\infty \left(\tilde{H}^{(1)}(t) - H^{(0)}(t)\right) f(t)\,dt,
\end{cases}
\tag{2.25}
$$

where, according to (1.10)–(1.13), $h^{(0)}(t) = \sum_{n=1}^{\infty} r * (v * r)^{*(n-1)}(t)$ is the density of 0-restoration function $H^{(0)}(t) = \sum_{n=1}^{\infty} R * (V * R)^{*(n-1)}(t)$ of alternating renewal process, generated by RV δ and γ; $h^{(1)}(t) = \sum_{n=1}^{\infty} (r * v)^{*(n)}(t)$ is the density of 1-restoration function $H^{(1)}(t) = \sum_{n=1}^{\infty} (R * V)^{*(n)}(t)$ of the same alternating renewal process; $\hat{H}^{(1)}(t)=1+H^{(1)}(t)$; $v^{(0)}(t,x)=\int_0^t v(x+t-y)h^{(1)}(y)dy$ is the DD of the control residual time; and $v^{(1)}(t,x)=r(t+x)+\int_0^t r(x+t-y)h^{(1)}(y)dy$ is the density of direct residual time till the control beginning.

The value of constant ρ_0 is obtained from the normalization requirement.

2.2.4 Stationary Characteristics Definition

Let us consider the phase state space E as a union of the following disjoint subsets:

$E_+ = \{111, 210x, 211x\}$ – system is in its up-state;

$E_- = \{100x, 222, 101x, 200\}$ – system is in its down-state.

Using (2.20), let us find average sojourn times in system states:

$$\begin{aligned}
&m(111)=\int_0^\infty \overline{F}(t)\overline{R}(t)dt,\ m(210x)=\int_0^x \overline{V}(t)dt,\\
&m(211x)=\int_0^x \overline{R}(t)dt,\ m(100x)=x,\\
&m(222)=E\beta,\ m(101x)=x,\ m(200)=E\gamma.
\end{aligned} \tag{2.26}$$

Let us determine average operating TF T_+ and average RT T_- with the help of formulas (1.17) and (1.18). First, we define expressions, included in (2.7), taking into account (2.23), (2.25), and (2.26).

$$\begin{aligned}
&\int_{E_+} m(e)\rho(de)=m(111)\rho(111)+\int_0^\infty m(210x)\rho(210x)dx+\\
&+\int_0^\infty m(211x)\rho(211x)dx=\\
&=\rho_0\left(\int_0^\infty \overline{F}(t)\overline{R}(t)dt+\int_0^\infty dx\int_0^x \overline{V}(t)dt\int_x^\infty f(y)h^{(0)}(y-x)dy+\right.\\
&\left.+\int_0^\infty dx\int_0^x \overline{R}(t)dt\int_x^\infty h^{(1)}(y-x)f(y)dy\right).
\end{aligned} \tag{2.27}$$

Let us transform the right part of the expression (2.27).

$$\int_0^\infty \overline{F}(t)\overline{R}(t)\,dt + \int_0^\infty dx\int_0^x \overline{V}(t)\,dt\int_x^\infty f(y)h^{(0)}(y-x)\,dy +$$

$$+\int_0^\infty dx\int_0^x \overline{R}(t)\,dt\int_x^\infty h^{(1)}(y-x)f(y)\,dy =$$

$$= \int_0^\infty f(y)\,dy\int_0^y \overline{R}(t)\,dt + \int_0^\infty dx\int_x^\infty f(y)h^{(0)}(y-x)\,dy\int_0^x \overline{V}(t)\,dt +$$

$$+\int_0^\infty dx\int_x^\infty h^{(1)}(y-x)f(y)\,dy\int_0^x \overline{R}(t)\,dt = \tag{2.28}$$

$$= \int_0^\infty f(y)\,dy\int_0^y \overline{R}(t)\,dt + \int_0^\infty f(y)\,dy\int_0^y h^{(0)}(y-x)\,dx\int_0^x \overline{V}(t)\,dt +$$

$$+\int_0^\infty f(y)\,dy\int_0^y h^{(1)}(y-x)\,dx\int_0^x \overline{R}(t)\,dt =$$

$$= \int_0^\infty f(y)\,dy\left(\int_0^y \overline{R}(t)\,dt + \int_0^y h^{(0)}(y-x)\,dx\int_0^x \overline{V}(t)\,dt + \int_0^y h^{(1)}(y-x)\,dx\int_0^x \overline{R}(t)\,dt\right).$$

Taking into account that

$$\int_0^y \overline{R}(t)dt + \int_0^y h^{(0)}(y-x)\,dx\int_0^x \overline{V}(t)\,dt + \int_0^y h^{(1)}(y-x)\,dx\int_0^x \overline{R}(t)\,dt = y,$$

the expression (2.28) will transform into

$$\int_0^\infty y\,f(y)\,dy = E\alpha.$$

Consequently,

$$\int_{E_+} m(e)\rho(de) = \rho_0 E\alpha. \tag{2.29}$$

Next,

$$
\begin{aligned}
&\int_{E_-} m(e)\rho(de) = m(222)\rho(222) + m(200)\rho(200) + \int_0^\infty m(100x)\rho(100x)\,dx + \\
&+\int_0^\infty m(101x)\rho(101x)\,dx = \rho_0\left(E\beta + E\gamma\int_0^\infty \left(\tilde{H}^{(1)}(t) - H^{(0)}(t)\right) f(t)\,dt + \right.\\
&\left. +\int_0^\infty x\,dx\int_0^\infty f(t)v^{(0)}(t,x)\,dt + \int_0^\infty x\,dx\int_0^\infty f(t)v^{(1)}(t,x)\,dt \right) = \\
&= \rho_0\left(E\beta + E\gamma\int_0^\infty \left(\tilde{H}^{(1)}(t) - H^{(0)}(t)\right) f(t)\,dt + \right.\\
&\left. +\int_0^\infty f(t)\,dt\int_0^\infty x\left(v^{(0)}(t,x) + v^{(1)}(t,x)\right)dx \right) = \\
&= \rho_0\left(E\beta + E\gamma\int_0^\infty \left(\tilde{H}^{(1)}(t) - H^{(0)}(t)\right) f(t)\,dt + \right.\\
&\left. +\int_0^\infty f(t)\,dt\int_0^\infty \left(\bar{V}^{(0)}(t,x) + \bar{V}^{(1)}(t,x)\right)dx \right) = \\
&= \rho_0\left(E\beta + E\gamma\int_0^\infty \left(\tilde{H}^{(1)}(t) - H^{(0)}(t)\right) f(t)\,dt + \right.\\
&\left. +E\gamma\int_0^\infty H^{(0)}(t)f(t)\,dt + E\delta\int_0^\infty \tilde{H}^{(1)}(t)f(t)\,dt - E\alpha \right) = \\
&= \rho_0\left(E\beta + (E\gamma + E\delta)\int_0^\infty \tilde{H}^{(1)}(t)f(t)\,dt - E\alpha \right).
\end{aligned}
\tag{2.30}
$$

To get (2.30), the following identity was used:

$$
\int_0^\infty \left(\bar{V}^{(0)}(t,x) + \bar{V}^{(1)}(t,x)\right)dx = E\gamma H^{(0)}(t) + E\delta\tilde{H}^{(1)}(t) - t.
$$

Then,

$$
\int_E m(e)\rho(de) = \rho_0\left(E\beta + \left(E\gamma + E\delta\right)\int_0^\infty \tilde{H}^{(1)}(t)f(t)\,dt \right).
$$

Next,

$$\begin{aligned}&\int_{E_+} P(e,E_-)\rho(de)=\rho(111)P(111,E_-)+\int_0^\infty \rho(210x)P(210x,E_-)\,dx+\\&+\int_0^\infty \rho(211x)P(211x,E_-)\,dx=\rho_0\left(\int_0^\infty dx\int_0^\infty f(t)r(t+x)\,dt+\right.\\&+\int_0^\infty dx\int_x^\infty h^{(0)}(t-x)f(t)\,dt\int_0^\infty v(x+y)\,dy+\\&\left.+\int_0^\infty dx\int_x^\infty h^{(1)}(t-x)f(t)\,dt\int_0^\infty r(x+y)\,dy\right).\end{aligned}\tag{2.31}$$

Let us simplify the first summand of the expression (2.31).

$$\begin{aligned}&\int_0^\infty dx\int_0^\infty f(t)r(t+x)\,dt=\int_0^\infty f(t)\,dt\int_0^\infty r(t+x)\,dx=\int_0^\infty f(t)\,dt\int_t^\infty r(x)\,dx=\\&=\int_0^\infty \overline{R}(t)f(t)\,dt.\end{aligned}\tag{2.32}$$

Transform the second summand of the expression (2.31).

$$\begin{aligned}&\int_0^\infty dx\int_x^\infty h^{(0)}(t-x)f(t)\,dt\int_0^\infty v(x+y)\,dy=\int_0^\infty dx\int_x^\infty h^{(0)}(t-x)f(t)\,dt\int_x^\infty v(y)\,dy=\\&=\int_0^\infty dx\int_x^\infty h^{(0)}(t-x)f(t)\overline{V}(x)\,dt=\int_0^\infty f(t)\,dt\int_0^t \overline{V}(x)h^{(0)}(t-x)\,dx.\end{aligned}\tag{2.33}$$

Similarly, the third summand of (2.31) will be

$$\int_0^\infty dx\int_x^\infty h^{(1)}(t-x)f(t)\,dt\int_0^\infty r(x+y)\,dy=\int_0^\infty f(t)\,dt\int_0^t \overline{R}(x)h^{(1)}(t-x)\,dx.\tag{2.34}$$

Therefore, taking into account formulas (2.32)–(2.34), the expression (2.31) can be rewritten as follows:

$$\begin{aligned}&\int_{E_+} P(x,E_-)\rho(de)=\rho_0\left(\int_0^\infty \overline{R}(t)f(t)\,dt+\int_0^\infty f(t)\,dt\int_0^t \overline{V}(x)h^{(0)}(t-x)\,dx+\right.\\&\left.+\int_0^\infty f(t)\,dt\int_0^t \overline{R}(x)h^{(1)}(t-x)\,dx\right)=\\&=\rho_0\int_0^\infty f(t)\,dt\left(\overline{R}(t)+\int_0^t \overline{V}(x)h^{(0)}(t-x)\,dx+\int_0^t \overline{R}(x)h^{(1)}(t-x)\,dx\right)=\\&=\rho_0\int_0^\infty f(t)\,dt=\rho_0.\end{aligned}\tag{2.35}$$

To transform (2.35), the following identity was used:

$$\bar{R}(t) + \int_0^t \bar{V}(x)h^{(0)}(t-x)\,dx + \int_0^t \bar{R}(x)h^{(1)}(t-x)\,dx = 1.$$

Therefore, average operating TF T_+ is:

$$T_+ = E\alpha.$$

Average stationary RT T_- is defined by the formula

$$T_- = E\beta - E\alpha + (E\gamma + E\delta)\int_0^\infty \tilde{H}^{(1)}(t)f(t)\,dt.$$

Then, the stationary availability factor K_a is

$$K_a = \frac{E\alpha}{E\beta + (E\delta + E\gamma)\int_0^\infty \tilde{H}^{(1)}(t)f(t)\,dt}. \tag{2.36}$$

Let us point out probability sense of the expression in (2.36):

$\int_0^\infty \tilde{H}^{(1)}(t)f(t)\,dt$ is an average number of controls before the latent failure occurrence.

Let us get the formula for K_a under nonrandom control periodicity $\tau > 0$. In this case, $R(t) = 1(t-\tau)$, $E\delta = \tau$, the expression (2.36) takes the following form:

$$K_a = \frac{E\alpha}{E\beta + (\tau + E\gamma)\left(1 + \sum_{n=1}^{\infty}\int_0^\infty \bar{F}(z + n\tau)v^{*(n)}dz\right)}. \tag{2.37}$$

Assume that both control periodicity and control duration are nonrandom: $\tau > 0$, $h > 0$. Taking into consideration that in this case $R(t) = 1(t-\tau)$, $E\delta = T$, $V(t) = 1(t-h)$, $E\gamma = h$, the formula (2.36) will be

$$K_a = \frac{E\alpha}{E\beta + (\tau + h)\sum_{n=0}^{\infty}\bar{F}(n(\tau + h))}. \tag{2.38}$$

Let us find stationary efficiency characteristics of the system: average specific income S per calendar time unit and average specific expenses C per time unit of up-state according to formulas (1.22) and (1.23), respectively.

Assume c_1 is the income per time unit of component up-state, c_2 are expenses per time unit of component restoration, c_3 are expenses per time unit of control execution, and c_4 are wastes per time unit of latent failure. Then, for the given system, the functions $f_s(e), f_c(e)$ are as follows:

$$f_s(e)=\begin{cases} c_1, e\in\{111,210x\},\\ c_1-c_3, e=211x,\\ -c_2, e=200,\\ -c_3-c_4, e\in\{101x,201\},\\ -c_4, e\in\{100x\},\end{cases}\qquad f_c(e)=\begin{cases} 0, e\in\{111,210x\},\\ c_2, e=200,\\ c_3, e=211x,\\ c_3+c_4, e\in\{101x,201\},\\ c_4, e=100x.\end{cases}\tag{2.39}$$

Using formulas (2.23), (2.25), (2.26), and (2.39), let us find the values of expressions, included in formulas (1.22) and (1.23):

$$\begin{aligned}
&\int_E m(e)f_s(e)\rho(de)=c_1 m(111)\rho(111)+(c_1-c_3)\int_0^\infty m(210x)\rho(210x)\,dx+\\
&+c_1\int_0^\infty m(211x)\rho(211x)\,dx-c_2 m(222)\rho(222)-\\
&-(c_3+c_4)m(200)\rho(200)-(c_3+c_4)\int_0^\infty m(100x)\rho(100x)\,dx-\\
&-c_4\int_0^\infty m(101x)\rho(101x)\,dx=\\
&=\rho_0\left(c_1E\alpha-c_2E\beta-c_4\left((E\gamma+E\delta)\int_0^\infty \tilde{H}^{(1)}(t)f(t)\,dt-E\alpha\right)-\right.\\
&-c_3\left(E\gamma\int_0^\infty\left(\tilde{H}^{(1)}(t)-H^{(0)}(t)\right)f(t)\,dt+\right.\\
&\left.\left.+\int_0^\infty f(t)\,dt\int_0^t h^{(0)}(t-x)\,dx\int_0^x \bar{V}(y)\,dy+\int_0^\infty f(t)\,dt\int_0^\infty \bar{V}^{(0)}(t,x)\,dx\right)\right)=\\
&=\rho_0\left((c_1+c_4)E\alpha-c_2E\beta-\left(c_3E\gamma+c_4(E\gamma+E\delta)\right)\int_0^\infty \tilde{H}^{(1)}(t)f(t)\,dt\right),
\end{aligned}\tag{2.40}$$

$$\begin{aligned}
&\int_E m(e)f_c(e)\rho(de) = c_2 m(222)\rho(222) + c_3\int_0^\infty m(210x)\rho(210x)\,dx + c_3 m(200)\rho(200) + \\
&+c_3\int_0^\infty m(100x)\rho(100x)\,dx + c_4\int_0^\infty m(100x)\rho(100x)\,dx + \\
&+c_4\int_0^\infty m(101x)\rho(101x)\,dx + c_4 m(200)\rho(200) = \\
&= \rho_0\left(c_4\left((E\gamma+E\delta)\int_0^\infty f(t)\tilde{H}^{(1)}(t)\,dt - E\alpha\right) + c_2 E\beta + \right. \\
&+c_3\left(\int_0^\infty f(t)\,dt\int_0^t h^{(0)}(t-x)\,dx\int_0^x \bar{V}(y)\,dy + \right. \\
&\left.\left.+\int_0^\infty f(t)\,dt\int_0^\infty \bar{V}^{(0)}(t,x)\,dx + E\gamma\int_0^\infty\left(\tilde{H}^{(1)}(t) - H^{(0)}(t)\right)f(t)\,dt\right)\right) = \\
&= \rho_0\left(c_2 E\beta - c_4 E\alpha + \left((c_3+c_4)E\gamma + c_4 E\delta\right)\int_0^\infty \tilde{H}^{(1)}(t)f(t)\,dt\right).
\end{aligned} \tag{2.41}$$

The following identity was applied to transform (2.40) and (2.41):

$$\begin{aligned}
&\int_0^\infty f(t)\,dt\int_0^t h^{(0)}(t-x)\,dx\int_0^x \bar{V}(y)\,dy + \int_0^\infty f(t)\,dt\int_0^\infty \bar{V}^{(0)}(t,x)\,dx = \\
&= E\gamma\int_0^\infty H^{(0)}(t)f(t)\,dt.
\end{aligned} \tag{2.42}$$

The proof of (2.42) is given in Appendix A.

Consequently, average specific income per unit of calendar time can be calculated with the help of the ratio:

$$S = \frac{(c_1+c_4)E\alpha - c_2 E\beta - \left(c_3 E\gamma + c_4(E\gamma+E\delta)\right)\int_0^\infty \tilde{H}^{(1)}(t)f(t)\,dt}{E\beta + \left(E\delta + E\gamma\right)\int_0^\infty \tilde{H}^{(1)}(t)f(t)\,dt}. \tag{2.43}$$

Average specific expenses per time unit of system up-state are determined by

$$C = c_2\frac{E\beta}{E\alpha} + \left((c_3+c_4)\frac{E\gamma}{E\alpha} + c_4\frac{E\delta}{E\alpha}\right)\int_0^\infty \tilde{H}^{(1)}(t)f(t)\,dt - c_4. \tag{2.44}$$

Let us get formulas for S and C under the condition of nonrandom control periodicity $\tau > 0$. Since $R(t) = 1(t-\tau)$, $E\delta = T$, the expressions (2.43) and (2.44) transform into the following, respectively:

$$S = \frac{(c_1 + c_4)E\alpha - c_2 E\beta - \left(c_3 E\gamma + c_4(E\gamma + \tau)\right)\left(1 + \sum_{n=1}^{\infty}\int_0^{\infty}\bar{F}(z+n\tau)v^{*(n)}dz\right)}{E\beta + (\tau + E\gamma)\left(1 + \sum_{n=1}^{\infty}\int_0^{\infty}\bar{F}(z+n\tau)v^{*(n)}dz\right)}, \quad (2.45)$$

$$C = c_2\frac{E\beta}{E\alpha} + \left((c_3 + c_4)\frac{E\gamma}{E\alpha} + c_4\frac{E\delta}{E\alpha}\right)\left(1 + \sum_{n=1}^{\infty}\int_0^{\infty}\bar{F}(t+n\tau)v^{*(n)}(t)dt\right) - c_4. \quad (2.46)$$

Now, system efficiency characteristics under nonrandom control periodicity $\tau > 0$ and nonrandom control duration $h > 0$ can be obtained. Taking into consideration that $R(t) = 1(t-\tau)$, $E\delta = T$, $V(t) = 1(t-h)$, $E\gamma = h$, the expressions (2.43) and (2.44) transform into the following, respectively:

$$S = \frac{(c_1 + c_4)E\alpha - c_2 E\beta - \left((c_3 + c_4)h + c_4\tau\right)\sum_{n=0}^{\infty}\bar{F}(n(\tau + h))}{E\beta + (\tau + h)\sum_{n=0}^{\infty}\bar{F}(n(\tau + h))}, \quad (2.47)$$

$$C = c_2\frac{E\beta}{E\alpha} + \left((c_3 + c_4)\frac{h}{E\alpha} + c_4\frac{\tau}{E\alpha}\right)\sum_{n=0}^{\infty}\bar{F}(n(\tau + h)) - c_4. \quad (2.48)$$

One should note, under $h = 0$ and $\tau \to 0$ we get the results for the system with continuous control.

Example. Initial data and results of calculation of $K_a(\tau)$, $S(\tau)$, and $C(\tau)$, according to (2.38), (2.47), and (2.48), are represented in Table 2.2, RV α has exponential distribution and Erlangian distribution of the 4th order. To calculate average income and average expenses, the following initial data are taken: $\mathbf{c}_1$ = 5 c.u.; $\mathbf{c}_2$ = 3 c.u.; $\mathbf{c}_3$ = 2 c.u.; $\mathbf{c}_4$ = 4 c.u.

TABLE 2.2 Values of $K_a(\tau)$, $S(\tau)$, and $C(\tau)$ Under τ = 5 h

Initial data				Results		
Kind of RV α distribution	$E\alpha$, h	$E\beta$, h	$E\gamma$, h	$K_a(\tau)$	$S(\tau)$, c.u./h	$C(\tau)$ c.u./h
Exponential	60	0.500	0.200	0.949	4.522	0.242
Erlangian of the 4th order	60	0.500	0.200	0.951	4.526	0.240

2.3 APPROXIMATION OF STATIONARY CHARACTERISTICS OF ONE-COMPONENT SYSTEM WITHOUT COMPONENT DEACTIVATION

2.3.1 System Description

Let us investigate the system S operating as described in Section 2.2. Time diagram of its operation is represented in Figure 2.2. To approximate the initial system S stationary characteristics, we apply the method, developed in Ref. [14] and described in Section 1.3.

Let us choose a supporting system $S^{(0)}$. Assume the component operating TF is greater than RT and control duration. Then supporting system $S^{(0)}$ corresponds immediate restoration and control.

2.3.2 Semi-Markov Model Building of the Supporting System

Time diagram of supporting system $S^{(0)}$ operation and its transition graph are represented in Figures 2.9 and 2.10, respectively, momentary states of $S^{(0)}$ are indicated in brackets. Supporting system is simpler than the initial system described in Section 2.2.

System $S^{(0)}$ sojourn times are given by

$$\theta_{111} = \alpha \wedge \delta, \theta_{210x} = 0, \theta_{211x} = \delta \wedge x, \theta_{100x} = 0, \theta_{222} = 0, \theta_{101x} = x, \theta_{200} = 0.$$

For the initial system, they are written out in Section 2.2.

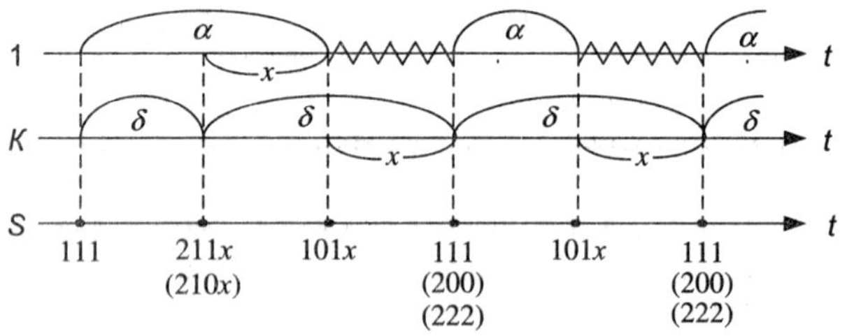

FIGURE 2.9 Time diagram of supporting system operation

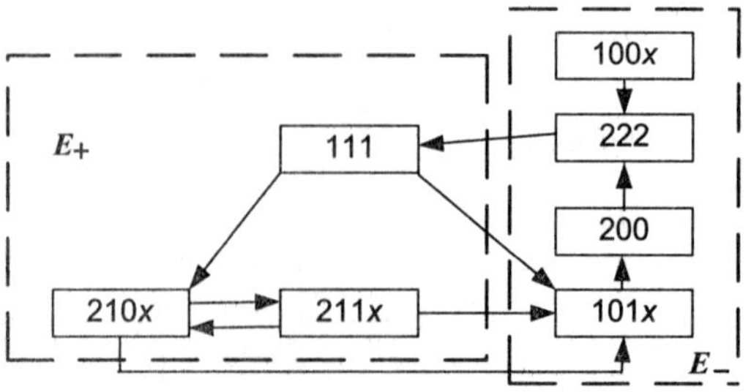

FIGURE 2.10 Supporting system transition graph

Let us describe transition events from the states of supporting system. Transition events from 111, 211x are illustrated in Figures 2.11 and 2.12, respectively, and are defined as follows:

$$\begin{aligned}&\{111 \to 210dx\} = \{\alpha - \delta \in dx\}, x > 0;\\&\{111 \to 101dx\} = \{\delta - \alpha \in dx\}, x > 0;\end{aligned} \tag{2.49}$$

$$\begin{aligned}&\{211x \to 210dy\} = \{x - \delta \in dy\}, 0 < y < x;\\&\{211x \to 101dy\} = \{\delta - x \in dy\}, y > 0.\end{aligned} \tag{2.50}$$

Transitions $210x \to 211x$, $100x \to 222$, $222 \to 111$, $101x \to 200$, and $200 \to 222$ occur with unity probability.

Taking into account formulas (2.49) and (2.50), let us define probabilities transitions of EMC $\{\xi_n^{(0)}; n \ge 0\}$ for supporting system.

$$\begin{aligned}
&p_{111}^{210dx} = P\{\alpha - \delta \in dx\} = \int_0^\infty f(x+t)r(t)dt\,dx, \quad x > 0;\\
&p_{111}^{101dx} = P\{\delta - \alpha \in dx\} = \int_0^\infty r(x+t)f(t)dtdx, \quad x > 0;\\
&p_{211x}^{210dy} = P\{x - \delta \in dy\} = r(x-y)dy, \quad 0 < y < x;\\
&p_{211x}^{101dy} = P\{\delta - x \in dy\} = r(y+x)dy, \quad y > 0;\\
&P_{210x}^{211x} = P_{100x}^{222} = P_{222}^{111} = P_{101x}^{200} = P_{200}^{222} = 1.
\end{aligned} \tag{2.51}$$

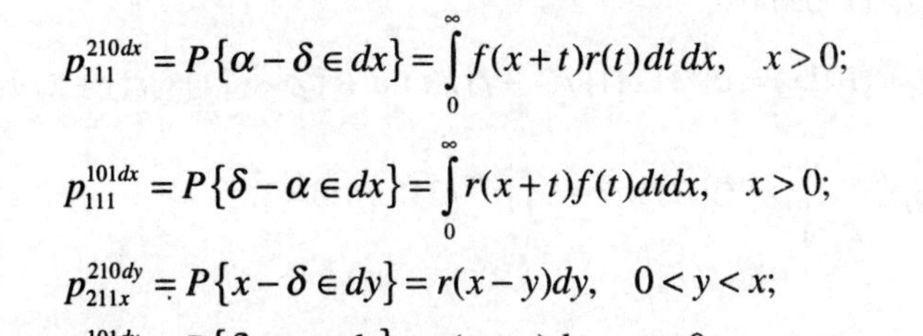

FIGURE 2.11 Transition events from the state 111

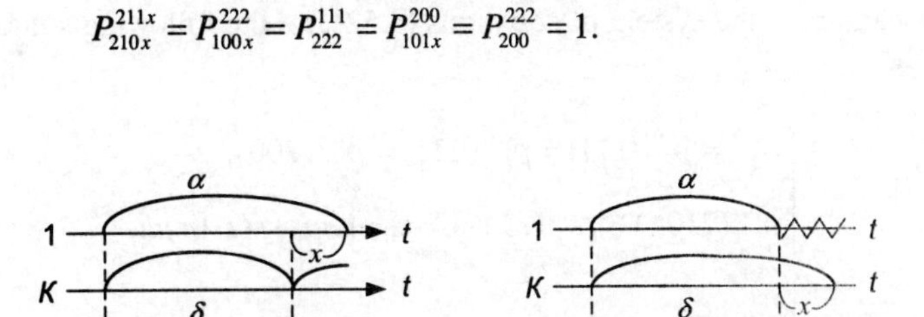

FIGURE 2.12 Transition events from the state 211x

2.3.3 Definition of EMC Stationary Distribution for Supporting System

Let us denote by $\rho^{(0)}(111)$, $\rho^{(0)}(222)$, and $\rho^{(0)}(200)$ the values of EMC $\{\xi_n^{(0)}; n \geq 0\}$ stationary distribution in states 111, 222, and 200, respectively, and assume the existence of stationary densities $\rho^{(0)}(210x)$, $\rho^{(0)}(211x)$, $\rho^{(0)}(100x)$, and $\rho^{(0)}(101x)$ for states 210x, 211x, 100x, and 101x, respectively. Let us construct the system of integral equations for them.

$$\begin{cases} \rho_0^{(0)} = \rho^{(0)}(111) = \rho^{(0)}(222) = \rho^{(0)}(200), \\ \rho^{(0)}(111) = \int\limits_0^\infty \rho^{(0)}(101x)dx, \\ \rho^{(0)}(210x) = \rho^{(0)}(111)\int\limits_0^\infty f(x+t)r(t)dt + \int\limits_x^\infty \rho^{(0)}(211y)r(y-x)dy, \\ \rho^{(0)}(211x) = \rho^{(0)}(210x), \\ \rho^{(0)}(100x) = 0, \\ \rho^{(0)}(101x) = \rho^{(0)}(111)\int\limits_0^\infty r(x+t)f(t)dt + \int\limits_0^\infty \rho^{(0)}(211y)r(y+x)dy, \\ 3\rho_0^{(0)} + \int\limits_0^\infty \rho^{(0)}(101x)dx + 2\int\limits_0^\infty \rho^{(0)}(210x)dx = 1. \end{cases} \tag{2.52}$$

The last equation in (2.52) is a normalization requirement.

One can prove the system of equations (2.52) has the following solution:

$$\begin{cases} \rho_0^{(0)} = \rho^{(0)}(111) = \rho^{(0)}(222) = \rho^{(0)}(200), \\ \rho^{(0)}(210x) = \rho^{(0)}(211x) = \rho_0^{(0)}\int\limits_0^\infty h_r(t)f(x+t)dt, \\ \rho^{(0)}(101x) = \rho_0^{(0)}\int\limits_0^\infty v_r(z,x)f(z)dz, \\ \rho^{(0)}(100x) = 0, \end{cases} \tag{2.53}$$

where $h_r(t) = \sum_{n=1}^\infty r^{*(n)}(t)$ is the density of renewal function $H_r(t)$ of renewal process generated by RV δ, and $v_r(z,x) = r(z+x) + \int\limits_0^z r(z+x-s)h_r(s)ds$ is the DD of the direct residual time for the same renewal process; the constant $\rho_0^{(0)}$ can be found from the normalization requirement.

2.3.4 Approximation of the System Stationary Characteristics

Let us determine approximate stationary characteristics of the system described in Section 2.2. The set $E^{(0)}$ of ergodic states of supporting system $S^{(0)}$ is: $E^{(0)} = \{111, 210x, 211x, 222, 101x, 200\}$. For the initial system S, the set of up E_+ and down E_- states are

$$E_+ = \{111, 210x, 211x\}, E_- = \{100x, 222, 101x, 200\}.$$

Transition probabilities of EMC $\{\xi_n; n \ge 0\}$ and average sojourn times of the initial system are given by formulas (2.23) and (2.26), respectively.

Approximate values of average stationary operating TF T_+ and average stationary RT T_- can be calculated by averages of formulas (1.25) and (1.26), respectively:

$$T_+ \approx \frac{\int_{E_+} m(e)\rho^{(0)}(de)}{\int_{E_+} P(e,E_-)\rho^{(0)}(de)}, T_- \approx \frac{\int_{E_-} m(e)\rho^{(0)}(de)}{\int_{E_+} P(e,E_-)\rho^{(0)}(de)}, \tag{2.54}$$

where $\rho^{(0)}(de)$ is the EMC $\{\xi_n^{(0)}; n \ge 0\}$ stationary distribution, $m(e)$ is the average values of initial system dwelling times in the state $e \in E$, and $P(e, E_-)$ is the probabilities transitions of EMC $\{\xi_n; n \ge 0\}$.

We obtain expressions from (2.54) by using (2.23), (2.26), and (2.53).

$$\begin{aligned}
&\int_{E_+} m(e)\rho^{(0)}(de) = m(111)\rho^{(0)}(111) + \int_0^\infty m(210x)\rho^{(0)}(210x)dx + \\
&+\int_0^\infty m(211x)\rho^{(0)}(211x)dx = \rho_0^{(0)}\left(\int_0^\infty \overline{F}(t)\overline{R}(t)dt + \right. \\
&\left. +\int_0^\infty dx \int_0^x \overline{R}(t)dt \int_0^\infty h_r(y)f(x+y)dy + \int_0^\infty dx \int_0^x \overline{V}(t)dt \int_0^\infty h_r(s)f(x+s)ds \right).
\end{aligned} \tag{2.55}$$

Transforming the second summand of (2.55), we get:

$$\begin{aligned}
&\int_0^\infty dx \int_0^x \overline{R}(t)dt \int_0^\infty h_r(y)f(x+y)dy = \int_0^\infty \overline{R}(t)dt \int_0^\infty \overline{F}(t+y)h_r(y)dy = \\
&= \int_0^\infty \overline{F}(y)dy \int_0^y \overline{R}(t)h_r(y-t)dt = \int_0^\infty \overline{F}(y)R(y)dy,
\end{aligned} \tag{2.56}$$

Here the formula (1.3) is applied.

Transforming the third summand of expression (2.55), we have:

$$\begin{aligned}
&\int_0^\infty dx \int_0^x \overline{V}(t)dt \int_0^\infty h_r(s)f(x+s)ds = \\
&= \int_0^\infty \overline{V}(t)dt \int_0^\infty \overline{F}(t+s)h_r(s)ds = \int_0^\infty \overline{V}(t)dt \int_0^\infty H_r(s)f(t+s)ds.
\end{aligned} \tag{2.57}$$

With regard to (2.56) and (2.57), the expression (2.55) turns into

$$\int_{E_+} m(e)\rho^{(0)}(de) = \rho_0^{(0)}\left(E\alpha + \int_0^\infty \overline{V}(t)\,dt \int_0^\infty H_r(s)f(t+s)\,ds\right). \quad (2.58)$$

Next,

$$\begin{aligned}
&\int_{E_-} m(e)\rho^{(0)}(de) = \int_0^\infty m(101x)\rho^{(0)}(101x)\,dx + m(200)\rho^{(0)}(200) + m(222)\rho^{(0)}(222) = \\
&= \rho_0^{(0)}\left(\int_0^\infty x\,dx \int_0^\infty f(t)v_r(t,x)\,dt + E\gamma + E\beta\right) = \\
&= \rho_0^{(0)}\left(\int_0^\infty f(t)\,dt \int_0^\infty x v_r(t,x)\,dx + E\gamma + E\beta\right) = \\
&= \rho_0^{(0)}\left(E\gamma + E\beta + \int_0^\infty f(z)\left(E\delta \tilde{H}_r(z) - z\right)dz\right) = \\
&= \rho_0^{(0)}\left(E\gamma + E\beta - E\alpha + E\delta \int_0^\infty \tilde{H}_r(z)f(z)\,dz\right).
\end{aligned} \quad (2.59)$$

where $H_r(t) = \sum_{n=1}^{\infty} R^{*(n)}(t)$ is renewal function, $\tilde{H}_r(t) = 1 + H_r(t)$.

$$\begin{aligned}
&\int_{E_+} P(e,E_-)\rho^{(0)}(de) = P(111,E_-)\rho^{(0)}(111) + \\
&+ \int_0^\infty P(210x,E_-)\rho^{(0)}(210x)\,dx + \int_0^\infty P(211x,E_-)\rho^{(0)}(211x)\,dx = \\
&= \rho_0^{(0)}\left(\int_0^\infty dx \int_0^\infty r(x+t)f(t)\,dt + \int_0^\infty dx \int_0^\infty h_r(z)f(x+z)\,dz \int_0^\infty v(x+y)\,dy + \right. \\
&\left. + \int_0^\infty dx \int_0^\infty h_r(z)f(x+z)\,dz \int_0^\infty r(x+y)\,dy\right) = \rho_0^{(0)}\left(1 + \int_0^\infty \overline{V}(t)\,dt \int_0^\infty h_r(z)f(t+z)\,dz\right).
\end{aligned} \quad (2.60)$$

Taking into account (2.58) and (2.60), we get the following approximate formula for average stationary operation TF T_+:

$$T_+ \approx \frac{E\alpha + \int_0^\infty \overline{V}(t)\,dt \int_0^\infty H_r(x)f(x+t)\,dx}{1 + \int_0^\infty \overline{V}(t)\,dt \int_0^\infty h_r(x)f(t+x)\,dx}.$$

With regard to (2.59) and (2.60), average stationary RT T_- is approximated by

$$T_- \approx \frac{E\gamma + E\beta - E\alpha + E\delta \int_0^\infty \tilde{H}_r(x) f(x)dx}{1 + \int_0^\infty \overline{V}(t)dt \int_0^\infty h_r(x) f(t+x)dx}.$$

Stationary availability factor is given by the ratio $K_a = \dfrac{T_+}{T_+ + T_-}$. Then,

$$K_a \approx \frac{E\alpha + \int_0^\infty \overline{V}(t)dt \int_0^\infty H_r(x) f(x+t)dx}{E\gamma + E\beta + E\delta \int_0^\infty \tilde{H}_r(x) f(x)dx + \int_0^\infty \overline{V}(t)dt \int_0^\infty H_r(x) f(x+t)dx}. \tag{2.61}$$

Now we can get approximate formula for K_a, under the condition of nonrandom control periodicity $\tau > 0$. In this case $R(t) = 1(t-\tau)$, $E\delta = T$, the expression (2.61) takes the form

$$K_a \approx \frac{E\alpha + \sum_{n=1}^{\infty} \int_0^\infty \overline{V}(t) \overline{F}(t+n\tau)dt}{E\beta + E\gamma + \tau \sum_{n=0}^{\infty} \overline{F}(n\tau) + \sum_{n=1}^{\infty} \int_0^\infty \overline{V}(t)\overline{F}(t+n\tau)dt}. \tag{2.62}$$

Let us find system approximate stationary characteristics of efficiency: average specific income S per unit of calendar time and average specific expenses C per time unit of system up-state; to do it we apply (2.24).

For the given system, the functions $f_s(e)$, $f_c(e)$ are

$$f_s(e) = \begin{cases} c_1, & e \in \{111, 211x\}, \\ c_1 - c_3, & e = 210x, \\ -c_2, & e = 222, \\ -c_3 - c_4, & e = 200, \\ -c_4, & e = 101x, \end{cases} \qquad f_c(e) = \begin{cases} 0, & e \in \{111, 211x\}, \\ c_2, & e = 222, \\ c_3, & e = 210x, \\ c_3 + c_4, & e = 200, \\ c_4, & e = 101x. \end{cases} \tag{2.63}$$

Here, c_1 is the income per time unit of initial system up-state, c_2 denotes expenses per time unit of restoration of initial system, c_3 are expenses per time unit of control, and c_4 are expenses per time unit of latent failure.

With the regard to (2.23), (2.24), (2.53), and (2.63), one can find the expressions from (1.27):

$$\int_E m(e)f_s(e)\rho^{(0)}(de) = c_1\rho^{(0)}(111)m(111) + (c_1 - c_3)\int_0^\infty \rho^{(0)}(210x)m(210x)\,dx +$$
$$+c_1\int_0^\infty \rho^{(0)}(211x)m(211x)\,dx - c_2\rho^{(0)}(222)m(222) -$$
$$-(c_3+c_4)\rho^{(0)}(200)m(200) - c_4\int_0^\infty \rho^{(0)}(101x)m(101x)\,dx =$$
$$= \rho_0^{(0)}\left(c_1\left(\int_0^\infty \bar{F}(t)\bar{R}(t)\,dt + \int_0^\infty dx\int_0^\infty h_r(y)f(x+y)\,dy\int_0^x \bar{R}(t)\,dt +\right.\right.$$
$$\left.+\int_0^\infty dx\int_0^\infty h_r(y)f(x+y)\,dy\int_0^x \bar{V}(t)\,dt\right) - c_2E\beta - \tag{2.64}$$
$$\left.-c_3\left(\int_0^\infty dx\int_0^\infty h_r(y)f(x+y)\,dy\int_0^x \bar{V}(t)\,dt + E\gamma\right) - c_4\left(\int_0^\infty x\,dx\int_0^\infty v_r(y,x)f(y)\,dy + E\gamma\right)\right) =$$
$$= \rho_0^{(0)}\left((c_1+c_4)E\alpha + (c_1-c_3)\int_0^\infty \bar{V}(t)\,dt\int_0^\infty H_r(y)f(t+y)\,dy -\right.$$
$$\left.-(c_3+c_4)E\gamma - c_2E\beta - c_4E\delta\int_0^\infty \tilde{H}_r(y)f(y)\,dy\right),$$

$$\int_E m(e)f_c(e)\rho^{(0)}(de) = c_2\rho^{(0)}(222)m(222) + c_3\int_0^\infty \rho^{(0)}(210x)m(210x)\,dx +$$
$$+(c_3+c_4)\rho^{(0)}(200)m(200) + c_4\int_0^\infty \rho^{(0)}(101x)m(101x)\,dx =$$
$$= \rho_0^{(0)}\left(c_2E\beta + c_3\int_0^\infty dx\int_0^\infty h_r(y)f(x+y)\,dy\int_0^x \bar{V}(t)\,dt +\right.$$
$$\left.+(c_3+c_4)E\gamma + c_4\int_0^\infty x\,dx\int_0^\infty v_r(y,x)f(y)\,dy\right) = \tag{2.65}$$
$$= \rho_0^{(0)}\left(c_2E\beta + (c_3+c_4)E\gamma + c_3\int_0^\infty \bar{V}(t)\,dt\int_0^\infty H_r(y)f(t+y)\,dy +\right.$$
$$\left.+c_4\left(E\delta\int_0^\infty \tilde{H}_r(y)f(y)\,dy - E\alpha\right)\right).$$

Consequently, average specific income per calendar time unit is approximately equal:

$$S \approx \left((c_1 + c_4)E\alpha - c_2 E\beta - (c_3 + c_4) + (c_1 - c_3)\int_0^\infty \bar{V}(t)dt \int_0^\infty H_r(x)f(x+t)dx - \right.$$
$$\left. -c_4 E\delta \int_0^\infty \tilde{H}_r(x)f(x)dx \right) \Bigg/ \left(E\gamma + E\beta + E\delta \int_0^\infty \tilde{H}_r(x)f(x)dx + \right. \tag{2.66}$$
$$\left. + \int_0^\infty \bar{V}(t)dt \int_0^\infty H_r(x)f(x+t)dx \right).$$

Average specific expenses per time unit of up-state are

$$C \approx \left(c_2 E\beta + (c_3 + c_4)E\gamma + c_3 \int_0^\infty \bar{V}(t)dt \int_0^\infty H_r(x)f(x+t)dx + \right.$$
$$\left. + c_4 \left(E\delta \int_0^\infty \tilde{H}_r(t)f(t)dt - E\alpha \right) \right) \Bigg/ \left(E\alpha + \int_0^\infty \bar{V}(t)dt \int_0^\infty H_r(x)f(x+t)dx \right). \tag{2.67}$$

We can make approximation of the initial system efficiency characteristics, under the condition of nonrandom control periodicity $\tau > 0$. In this case $R(t) = 1(t - \tau)$, $E\delta = \tau$, the expressions (2.66) and (2.67) take the form

$$S(\tau) \approx \left((c_1 + c_4)E\alpha - c_2 E\beta - (c_3 + c_4)E\gamma + (c_1 - c_3)\sum_{n=1}^\infty \int_0^\infty \bar{V}(t)\bar{F}(t + n\tau)dt - \right.$$
$$\left. -c_4 \tau \sum_{n=0}^\infty \bar{F}(n\tau) \right) \Bigg/ \left(E\beta + E\gamma + \tau \sum_{n=0}^\infty \bar{F}(n\tau) + \sum_{n=1}^\infty \int_0^\infty \bar{V}(t)\bar{F}(t + n\tau)dt \right), \tag{2.68}$$

$$C(\tau) \approx \left(c_2 E\beta - c_4 E\alpha + (c_3 + c_4)E\gamma + c_3 \sum_{n=1}^\infty \int_0^\infty \bar{V}(t)\bar{F}(t + n\tau)dt + \right.$$
$$\left. + c_4 \tau \sum_{n=0}^\infty \bar{F}(n\tau) \right) \Bigg/ \left(E\alpha + \sum_{n=1}^\infty \int_0^\infty \bar{V}(t)\bar{F}(t + n\tau)dt \right). \tag{2.69}$$

Example. To calculate stationary availability factor, average specific income and average specific expenses with the help of approximate formulas (2.62), (2.68), and (2.69), let us use initial data and calculation results from Section 2.2. Initial data and calculation results of $K_a(\tau)$, $S(\tau)$, $C(\tau)$ are given in Table 2.3.

TABLE 2.3 Approximate Values of $K_a(\tau)$, $S(\tau)$, and $C(\tau)$ Under $\tau = 5$ h

Initial data				Results		
Kind of distribution of RV α	$E\alpha$, h	$E\beta$, h	$E\gamma$, h	$K_a(\tau)$	$S(\tau)$, c.u./h	$C(\tau)$ c.u./h
Exponential	60	0.500	0.200	0.956	4.501	0.263
Erlangian of the 4th order	60	0.500	0.200	0.964	4.502	0.264

TABLE 2.4 Comparison of Exact and Approximate Results

Kind of distribution of RV α	K_a, calculated by (2.38)	K_a, calculated by (2.62)	Error (%)
Exponential	0.949	0.956	0.72
Erlangian of the 4th order	0.951	0.964	1.30

By means of Tables 2.2 and 2.3, one can estimate the accuracy of calculation using approximate formulas. The comparative results are given in Table 2.4.

2.4 THE SYSTEM MODEL WITH COMPONENT DEACTIVATION AND POSSIBILITY OF CONTROL ERRORS

2.4.1 System Description

The system S consisting of an operable component and control equipment is under consideration. At the time zero, the system begins operating and the control is on. System TF is RV α with DF $F(t) = P\{\alpha \le t\}$ and DD $f(t)$. The control is executed in random time δ with DF $R(t) = P\{\delta \le t\}$ and DD $r(t)$. Control duration is RV γ with DF $V(t) = P\{\gamma \le t\}$ and DD $v(t)$. While control execution, the component is deactivated. In of case failure detection (all failures are latent, that is detected after control execution only), the component restoration begins. Component RT is RV β with DF $G(t) = P\{\beta \le t\}$ and DD $g(t)$. While restoration, the control is off. After restoration, all component properties are restored.

Since control equipment is supposed to be nonreliable, the errors of first and second orders are possible. An error of first order means the operable component is treated as failed. Its possibility is P_1. An error of second order is the failure omission. The possibility of such an error is P_0. RV α, β, δ, γ are assumed to be independent and to have finite expectations.

2.4.2 Semi-Markov Model Building

To describe the system operation, let us use SMP $\xi(t)$ with discrete continuous state space. The set E of semi-Markov states:

$$E = \{111, 2\hat{1}0x, 211x, 101x, 2\hat{0}0, 2\bar{0}1, 222, 2\bar{1}2\}.$$

For the given system, the meaning of semi-Markov state codes is

111 – the system has been restored, the control is on;
$2\hat{1}0x$ – control has begun, the system is in up-state, it does not operate during control execution, time $x > 0$ is left till the latent failure (regardless of control execution time);
$211x$ – control has ended; operable system component is treated correctly and continues to operate, time $x > 0$ is left till the latent failure (regardless of control execution time);
$101x$ – latent failure has occurred, time $x > 0$ is left till control execution;
$2\hat{0}0$ – control has begun, the system is in a state of latent failure, it does not operate;
$2\bar{0}1$ – control has ended, failed system is regarded as operable, its operation has begun;
222 – control has ended, failed system component is treated correctly, system restoration has begun, control is deactivated;
$2\bar{1}2$ – control has ended, system up-state is taken for failed one, system restoration has begun, control is deactivated;

Time diagram and transitions graph of the system are in Figures 2.13 and 2.14. Sojourn times in states are:

$$\begin{aligned}&\theta_{111} = \alpha \wedge \delta, \theta_{2\hat{1}0x} = \gamma, \theta_{211x} = x \wedge \delta, \theta_{101x} = x,\\&\theta_{2\hat{0}0} = \gamma, \theta_{2\bar{0}1} = \delta, \theta_{222} = \beta, \theta_{2\bar{1}2} = \beta.\end{aligned} \tag{2.70}$$

Let us describe system transitions. Introduce RV χ_0 and χ_1, so that $P(\chi_0 = 0) = p_0$, $P(\chi_0 = 1) = \bar{p}_0$, $\bar{p}_0 = 1 - p_0$, $P(\chi_1 = 0) = p_1$, $P(\chi_1 = 1) = \bar{p}_1$, $\bar{p}_1 = 1 - p_1$. Transition events from the states 212*x*, 202 are given in Figures 2.15 and 2.16, respectively, and are defined by the expressions:

$$\{2\hat{1}0x \to 211x\} = \{\chi_1 = 1\}, \quad \{2\hat{1}0x \to 2\bar{1}2\} = \{\chi_1 = 0\}; \tag{2.71}$$

$$\{2\hat{0}0 \to 2\bar{0}1\} = \{\chi_0 = 0\}, \quad \{2\hat{0}0 \to 222\} = \{\chi_0 = 1\}. \tag{2.72}$$

Transitions from the states 111, 211*x* are defined similarly to transitions considered in Section 2.1. Transitions $101x \to 2\hat{0}0$, $2\bar{0}1 \to 2\hat{0}0$, $222 \to 111$, and $2\bar{1}2 \to 111$ occur with the unity probability.

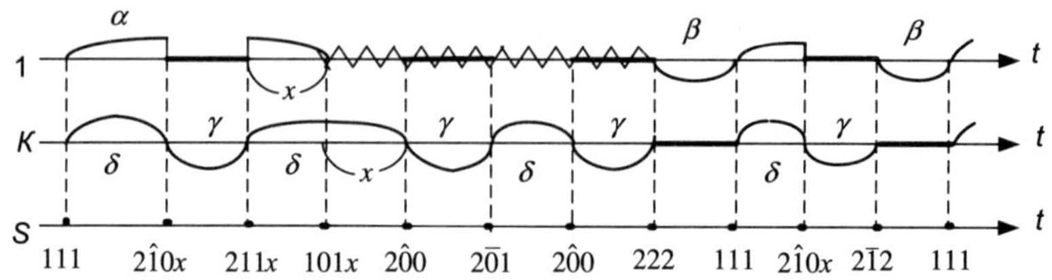

FIGURE 2.13 Time diagram of system operation

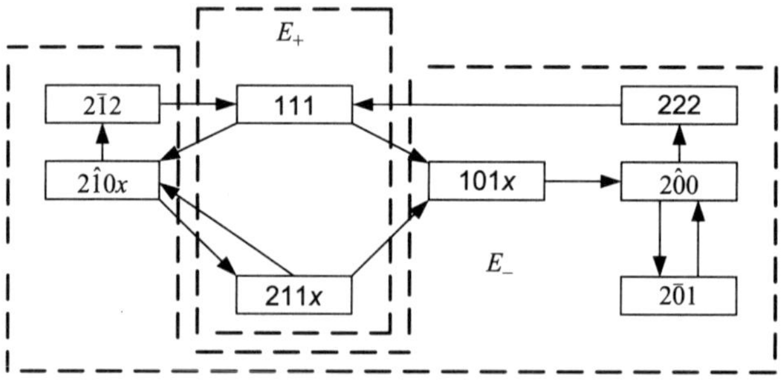

FIGURE 2.14 System transition graph

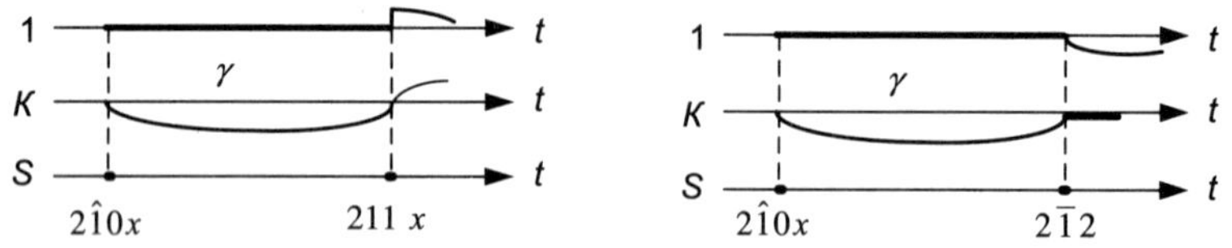

FIGURE 2.15 Transitions from 2î0x

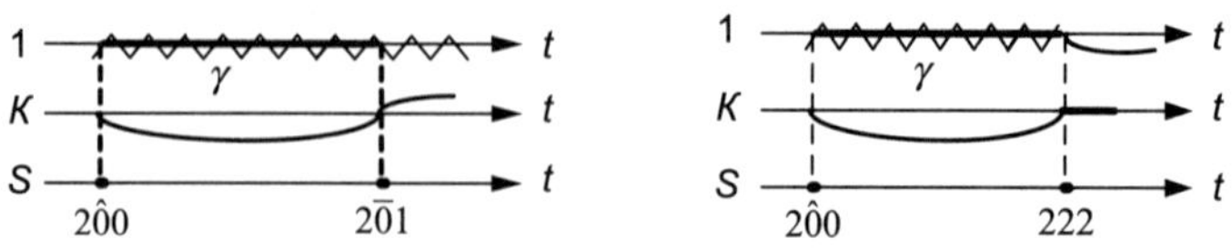

FIGURE 2.16 Transitions from 2ô0

Formulas (2.71) and (2.72) allow to obtain EMC $\{\xi_n; n \geq 0\}$ transition probabilities:

$$\begin{aligned}
&p_{111}^{2\hat{1}0dx} = \int_0^\infty f(x+t)r(t)dt\,dx, x>0; p_{111}^{101dx} = \int_0^\infty r(x+t)f(t)dtdx, x>0;\\
&p_{211x}^{101dy} = r(x+y)dy, y>0; p_{211x}^{2\hat{1}0dy} = r(x-y)dy, 0<y<x;\\
&P_{2\hat{1}0x}^{211x} = \overline{p}_1; P_{2\hat{1}0x}^{2\overline{1}2} = p_1; P_{2\hat{0}0}^{2\overline{0}1} = p_0; P_{2\hat{0}0}^{222} = \overline{p}_0;\\
&P_{101x}^{2\hat{0}0} = P_{2\overline{0}1}^{2\hat{0}0} = P_{222}^{111} = P_{2\overline{1}2}^{111} = 1.
\end{aligned} \tag{2.73}$$

2.4.3 Definition of EMC Stationary Distribution

Denote by $\rho(111)$, $\rho(2\hat{0}0)$, $\rho(2\overline{0}1)$, $\rho(222)$, $\rho(2\overline{1}2)$ the values of EMC $\{\xi_n; n \geq 0\}$ stationary distribution in the states 111, $2\hat{0}0$, $2\overline{0}1$, 222, $2\overline{1}2$ and assume the existence of stationary densities $\rho(2\hat{1}0x)$, $\rho(211x)$, $\rho(101x)$ for the states $2\hat{1}0x$, $211x$, $101x$, respectively.

The system of integral equations for the EMC $\{\xi_n;\, n \geq 0\}$ stationary distribution definition looks like the following:

$$\begin{cases}
\rho_0 = \rho(111) = \rho(222) + \rho(2\overline{1}2),\\
\rho(2\hat{1}0x) = \rho(111)\int_0^\infty f(x+t)r(t)dt + \int_x^\infty \rho(211y)r(y-x)dy,\\
\rho(211x) = \overline{p}_1\rho(2\hat{1}0x),\\
\rho(101x) = \rho(111)\int_0^\infty f(t)r(x+t)dt + \int_0^\infty \rho(211y)r(y+x)dy,\\
\rho(2\hat{0}0) = \int_0^\infty \rho(101x)dx + \rho(2\overline{0}1),\\
\rho(2\overline{0}1) = p_0\rho(2\hat{0}0),\\
\rho(222) = \overline{p}_0\rho(2\hat{0}0),\\
\rho(2\overline{1}2) = p_1\int_0^\infty \rho(2\hat{1}0x)dx,\\
2\rho_0 + \rho(2\hat{0}0) + \rho(2\overline{0}1) + \int_0^\infty \rho(2\hat{1}0x)dx + \int_0^\infty \rho(211x)dx + \int_0^\infty \rho(101x)dx = 1.
\end{cases} \tag{2.74}$$

The last equation in (2.74) is a normalization requirement.

One can prove the system of equations (2.74) has the following solution (Appendix B):

$$\begin{cases} \rho(2\hat{1}0x)=\dfrac{\rho_0}{\overline{p}_1}\int\limits_0^\infty h_{\tilde{r}}(y)f(x+y)dy, \\ \rho(211x)=\rho_0\int\limits_0^\infty h_{\tilde{r}}(y)f(x+y)dy, \\ \rho(101x)=\dfrac{\rho_0}{\overline{p}_1}\int\limits_0^\infty \tilde{v}(z,x)f(z)dz, \\ \rho(2\hat{0}0)=\rho_0\dfrac{1}{\overline{p}_0}\left(1-\dfrac{p_1}{\overline{p}_1}\int\limits_0^\infty H_{\tilde{r}}(z)f(z)dz\right), \\ \rho(2\overline{0}1)=\rho_0\dfrac{p_0}{\overline{p}_0}\left(1-\dfrac{p_1}{\overline{p}_1}\int\limits_0^\infty H_{\tilde{r}}(z)f(z)dz\right), \\ \rho(222)=\rho_0\left(1-\dfrac{p_1}{\overline{p}_1}\int\limits_0^\infty H_{\tilde{r}}(z)f(z)dz\right), \\ \rho(2\overline{1}2)=\rho_0\dfrac{p_1}{\overline{p}_1}\int\limits_0^\infty \overline{F}(y)h_{\tilde{r}}(y)dy. \end{cases} \tag{2.75}$$

Here $h_{\tilde{r}}(t)=\sum\limits_{n=1}^{\infty}\tilde{r}^{*(n)}(t)$ is the density of renewal function $H_{\tilde{r}}(t)=\sum\limits_{n=0}^{\infty}\tilde{R}^{*(n)}(t)$, generated by the improper DF $\tilde{R}(t)=\overline{p}_1R(t)$, $\tilde{r}(x)=\overline{p}_1r(x)$; $\tilde{r}^{*(n)}(t)$, $\tilde{R}^{*(n)}(t)$ is the n-fold convolutions of $\tilde{r}(t)$, $\tilde{R}(t)$.

The constant ρ_0 is found by means of normalization requirement.

2.4.4 System Stationary Characteristics Definition

The system phase state space E can be represented as a union of the following disjoint subsets:

$E_+ = \{111,\ 211x\}$ – system up-state;

$E_- = \{2\hat{1}0x,\ 101x,\ 2\hat{0}0,\ 2\overline{0}1,\ 222,\ 2\overline{1}2\}$ – system down-state.

Let us define average sojourn times in states with regard to (2.70):

$$\begin{gathered} m(111)=\int\limits_0^\infty \overline{F}(t)\overline{R}(t)dt, m(2\hat{1}0x)=E\gamma, m(211x)=\int\limits_0^x \overline{R}(t)dt, m(101x)=x, \\ m(2\hat{0}0)=E\gamma, m(2\overline{0}1)=E\delta, m(222)=E\beta, m(2\overline{1}2)=E\beta. \end{gathered} \tag{2.76}$$

Average stationary operating TF T_+ and average RT T_- can be determined by (1.17) and (1.18) with regard to expressions (2.75) and (2.76).

$$
\begin{aligned}
&\int_{E_+} m(e)\rho(de) = m(111)\rho(111) + \int_0^\infty m(211x)\rho(211x)dx = \\
&= \rho_0\left(\int_0^\infty \bar{F}(t)\bar{R}(t)dt + \int_0^\infty dx\int_0^x \bar{R}(t)dt\int_0^\infty h_{\bar{r}}(y)f(x+y)dy\right) = \\
&= \rho_0\left(\int_0^\infty \bar{F}(t)\bar{R}(t)dt + \int_0^\infty \bar{R}(t)dt\int_t^\infty dx\int_0^\infty h_{\bar{r}}(y)f(x+y)dy\right) = \\
&= \rho_0\left(\int_0^\infty \bar{F}(t)\bar{R}(t)dt + \int_0^\infty \bar{R}(t)dt\int_0^\infty h_{\bar{r}}(y)dy\int_t^\infty f(x+y)dx\right) = \\
&= \rho_0\left(\int_0^\infty \bar{F}(t)\bar{R}(t)dt + \int_0^\infty \bar{R}(t)dt\int_0^\infty h_{\bar{r}}(y)\bar{F}(y+t)dy\right) = \\
&= \rho_0\left(\int_0^\infty \bar{F}(t)\bar{R}(t)dt + \int_0^\infty \bar{R}(t)dt\int_t^\infty h_{\bar{r}}(y-t)\bar{F}(y)dy\right) = \\
&= \rho_0\left(\int_0^\infty \bar{F}(t)\bar{R}(t)dt + \int_0^\infty \bar{F}(y)dy\int_0^y h_{\bar{r}}(y-t)\bar{R}(t)dt\right) = \\
&= \rho_0\left(\int_0^\infty \bar{F}(t)\bar{R}(t)dt + \int_0^\infty \bar{F}(y)R(y)dy - \frac{p_1}{\bar{p}_1}\int_0^\infty H_{\bar{r}}(y)\bar{F}(y)dy\right) = \\
&= \rho_0\left(E\alpha - \frac{p_1}{\bar{p}_1}\int_0^\infty H_{\bar{r}}(y)\bar{F}(y)dy\right).
\end{aligned}
\tag{2.77}
$$

To simplify (2.77) the equality from Appendix B:

$$
\int_0^z \bar{R}(z-t)h_{\bar{r}}(t)dt = R(z) - \frac{p_1}{\bar{p}_1}H_{\bar{r}}(z)
$$

is applied.

Let us get,

$$\int_{E_-} m(e)\rho(de) = \int_0^\infty m(2\hat{1}0x)\rho(2\hat{1}0x)dx + \int_0^\infty m(101x)\rho(101x)dx + m(2\hat{0}0)\rho(2\hat{0}0) +$$
$$+ m(2\bar{0}1)\rho(2\bar{0}1) + m(222)\rho(222) + m(2\bar{1}2)\rho(2\bar{1}2) =$$
$$= \rho_0 \left(E\gamma \frac{1}{\bar{p}_1} \int_0^\infty dx \int_0^\infty h_{\bar{r}}(y) f(x+y)dy + \frac{1}{\bar{p}_1} \int_0^\infty x\,dx \int_0^\infty \tilde{v}(z,x) f(z)dz + \right.$$
$$+ E\gamma \frac{1}{\bar{p}_0} \left(1 - \frac{p_1}{\bar{p}_1} \int_0^\infty H_{\bar{r}}(z) f(z)dz \right) + E\delta \frac{p_0}{\bar{p}_0} \left(1 - \frac{p_1}{\bar{p}_1} \int_0^\infty H_{\bar{r}}(z) f(z)dz \right) +$$
$$\left. + E\beta \left(1 - \frac{p_1}{\bar{p}_1} \int_0^\infty H_{\bar{r}}(z) f(z)dz \right) + E\beta \frac{p_1}{\bar{p}_1} \int_0^\infty h_{\bar{r}}(z) \bar{F}(z)dz \right) = \tag{2.78}$$
$$= \rho_0 \left(E\gamma \frac{1}{\bar{p}_1} \int_0^\infty h_{\bar{r}}(y) \bar{F}(y)dy + \frac{1}{\bar{p}_1} \int_0^\infty f(z)dz \int_0^\infty x\tilde{v}(z,x)dx + \right.$$
$$\left. + E\gamma \frac{1}{\bar{p}_0} \left(1 - \frac{p_1}{\bar{p}_1} \int_0^\infty H_{\bar{r}}(z) f(z)dz \right) + E\delta \frac{p_0}{\bar{p}_0} \left(1 - \frac{p_1}{\bar{p}_1} \int_0^\infty H_{\bar{r}}(z) f(z)dz \right) + E\beta \right) =$$
$$= \rho_0 \left(E\gamma \int_0^\infty h_{\bar{r}}(z) \bar{F}(z)dz + E\delta \frac{\bar{p}_0 - p_1}{\bar{p}_1} \int_0^\infty h_{\bar{r}}(z) \bar{F}(z)dz + \right.$$
$$\left. + \frac{1}{\bar{p}_0} E\delta + \frac{1}{\bar{p}_0} E\gamma - E\alpha + \frac{p_1}{\bar{p}_1} \int_0^\infty H_{\bar{r}}(x) \bar{F}(x)dx + E\beta \right).$$

While transformations of (2.78), the following formula (Appendix B) was used:

$$\int_0^\infty x\tilde{v}(z,x)dx = \bar{p}_1 \int_z^\infty \bar{R}(x)dx + \bar{p}_1 \;\; E\gamma \;\; H_{\bar{r}}(z) + p_1 = \int_0^z H_{\bar{r}}(x)dx - \bar{p}_1 z + \bar{p}_1 \int_0^z \bar{R}(x)dx =$$
$$= \bar{p}_1 \;\; E\gamma + \bar{p}_1 \;\; E\gamma \;\; H_{\bar{r}}(z) + p_1 \int_0^z H_{\bar{r}}(x)dx - \bar{p}_1 z.$$

Next,

$$
\begin{aligned}
&\int_{E_+} P(e,E_-)\rho(de) = P(111,E_-)\rho(111) + \int_0^\infty P(211x,E_-)\rho(211x)dx = \\
&= \rho_0\left(\int_0^\infty dx\int_0^\infty f(x+t)r(t)dt + \int_0^\infty dx\int_0^\infty r(x+t)f(t)dt + \right.\\
&+\int_0^\infty dx\int_0^\infty f(x+z)h_{\bar r}(z)dz\int_0^\infty r(x+y)dy + \\
&\left.+\int_0^\infty dx\int_0^\infty f(x+z)h_{\bar r}(z)dz\int_0^x r(x-y)dy\right) = \\
&= \rho_0\left(\int_0^\infty r(t)dt\int_0^\infty f(x+t)dx + \int_0^\infty f(t)dt\int_0^\infty r(x+t)dx + \right. \qquad (2.79)\\
&\left.+\int_0^\infty dx\int_0^\infty f(x+z)h_{\bar r}(z)\bar R(x)dz + \int_0^\infty dx\int_0^\infty f(x+z)h_{\bar r}(z)R(x)dz\right) = \\
&= \rho_0\left(\int_0^\infty \bar F(t)r(t)dt + \int_0^\infty f(t)\bar R(t)dt + \int_0^\infty dx\int_0^\infty f(x+z)h_{\bar r}(z)dz\right) = \\
&= \rho_0\left(1+\int_0^\infty \bar F(z)h_{\bar r}(z)dz\right) = \rho_0\left(1+\int_0^\infty f(z)H_{\bar r}(z)dz\right).
\end{aligned}
$$

In such a way, average stationary operating TF T_+ is

$$
T_+ = \frac{E\alpha - \dfrac{p_1}{\bar p_1}\displaystyle\int_0^\infty \bar F(y)H_{\bar r}(y)dy}{1+\displaystyle\int_0^\infty \bar F(z)dH_{\bar r}(z)}. \qquad (2.80)
$$

Average stationary RT T_- can be calculated by the ratio:

$$
\begin{aligned}
T_- = &\left((E\delta + E\gamma)\left(\frac{\bar p_0 - p_1}{\bar p_0\bar p_1}\int_0^\infty \bar F(z)dH_{\bar r}(z) + \frac{1}{\bar p_0}\right) + \right.\\
&\left. + E\beta - E\alpha + \frac{p_1}{\bar p_1}\int_0^\infty H_{\bar r}(t)\bar F(t)dt\right) \Big/ \qquad (2.81)\\
&\Big/\left(1+\int_0^\infty \bar F(z)dH_{\bar r}(z)\right),
\end{aligned}
$$

and for the stationary availability factor K_a we get:

$$K_a = \frac{E\alpha - \frac{p_1}{\overline{p}_1}\int\limits_0^\infty \overline{F}(y)H_{\overline{r}}(y)dy}{(E\delta + E\gamma)\frac{\overline{p}_0 - p_1}{\overline{p}_0\overline{p}_1}\int\limits_0^\infty \overline{F}(z)dH_{\overline{r}}(z) + \frac{1}{\overline{p}_0}(E\delta + E\gamma) + E\beta}. \tag{2.82}$$

Let us define stationary efficiency characteristics: average specific income per calendar time unit S and average specific expenses per time unit of up-state C.

For the given system, the functions $f_s(e), f_c(e)$ look like the following:

$$f_s(e) = \begin{cases} c_1, & e \in \{111,\ \ 211x\}, \\ -c_2, & e \in \{222,\ \ 2\overline{1}2\}, \\ -c_3, & e \in \{2\hat{1}0x,\ \ 2\hat{0}0\}, \\ -c_4, & e \in \{101x,\ \ 2\overline{0}1\}, \end{cases} \qquad f_c(e) = \begin{cases} 0, & e \in \{111,\ \ 211x\}, \\ c_2, & e \in \{222,\ \ 2\overline{1}2\}, \\ c_3, & e \in \{2\hat{1}0x,\ \ 2\hat{0}0\}, \\ c_4, & e \in \{101x,\ \ 2\overline{0}1\}. \end{cases} \tag{2.83}$$

Here, c_1 is the income per time unit of component up-state, c_2 are expenses per time unit of component restoration, c_3 are expenses per time unit of control execution, and c_4 are expenses per time unit of latent failure.

With regard to (2.83), average specific income is

$$S = \Bigg((c_1 + c_4)\left(E\alpha - \frac{p_1}{\overline{p}_1}\int\limits_0^\infty \overline{F}(t)H_{\overline{r}}(t)dt \right) - c_2E\beta -$$
$$-(c_3E\gamma + c_4E\delta)\left(\frac{1}{\overline{p}_0} + \frac{\overline{p}_0 - p_1}{\overline{p}_0\overline{p}_1} \right)\int\limits_0^\infty \overline{F}(z)dH_{\overline{r}}(z) \Bigg) \Bigg/ \Bigg(\frac{1}{\overline{p}_0}(E\gamma + E\delta) + \tag{2.84}$$
$$+(E\gamma + E\delta)\frac{\overline{p}_0 - p_1}{\overline{p}_0\overline{p}_1}\int\limits_0^\infty \overline{F}(z)dH_{\overline{r}}(z) + E\beta \Bigg).$$

Average specific expenses are

$$C = \Bigg(c_2E\beta - c_4E\alpha + (c_4E\delta + c_3E\gamma)\frac{\overline{p}_0 - p_1}{\overline{p}_0\overline{p}_1}\int\limits_0^\infty \overline{F}(z)dH_{\overline{r}}(z) +$$
$$+c_4\frac{p_1}{\overline{p}_1}\int\limits_0^\infty \overline{F}(t)H_{\overline{r}}(t)dt + \frac{1}{\overline{p}_0}(c_3E\gamma + c_4E\delta) \Bigg) \Bigg/ \left(E\alpha - \frac{p_1}{\overline{p}_1}\int\limits_0^\infty \overline{F}(t)H_{\overline{r}}(t)dt \right). \tag{2.85}$$

Let us consider the specific case when $R(t) = 1(t - \tau),\ E\delta = \tau$.

Then average stationary operating TF T_+ and average stationary RT T_- are determined by the formulas:

$$T_+ = \frac{E\alpha - \sum_{n=1}^{\infty}(1-\overline{p}_1^n)\int_{n\tau}^{(n+1)\tau}\overline{F}(t)\,dt}{1+\sum_{n=1}^{\infty}\overline{p}_1^n\overline{F}(n\tau)} = \frac{\int_0^{\tau}\overline{F}(t)\,dt + \sum_{n=1}^{\infty}\overline{p}_1^n\int_{n\tau}^{(n+1)\tau}\overline{F}(t)\,dt}{1+\sum_{n=1}^{\infty}\overline{p}_1^n\overline{F}(n\tau)}, \tag{2.86}$$

$$T_- = \left((E\delta + E\gamma)\frac{\overline{p}_0 - p_1}{\overline{p}_0\overline{p}_1}\sum_{n=1}^{\infty}\overline{p}_1^n\overline{F}(n\tau) + \frac{1}{\overline{p}_0}(E\delta + E\gamma) + E\beta - E\alpha + \right.$$
$$\left. + \sum_{n=1}^{\infty}(1-\overline{p}_1^n)\int_{n\tau}^{(n+1)\tau}\overline{F}(t)\,dt\right)\Bigg/\left(1+\sum_{n=1}^{\infty}\overline{p}_1^n\overline{F}(n\tau)\right). \tag{2.87}$$

Stationary availability factor is as follows:

$$K_a = \frac{E\alpha - \sum_{n=1}^{\infty}(1-\overline{p}_1^n)\int_{n\tau}^{(n+1)\tau}\overline{F}(t)\,dt}{E\beta + \frac{1}{\overline{p}_0}(E\gamma+\tau) + (E\gamma+\tau)\frac{\overline{p}_0 - p_1}{\overline{p}_0\overline{p}_1}\sum_{n=1}^{\infty}\overline{p}_1^n\overline{F}(n\tau)}. \tag{2.88}$$

Average specific income and expenses are

$$S(\tau) = \left(E\alpha(c_1 + c_4) - c_2E\beta - (c_3E\gamma + c_4\tau)\frac{\overline{p}_0 - p_1}{\overline{p}_0\overline{p}_1}\sum_{n=1}^{\infty}\overline{p}_1^n\overline{F}(n\tau) - \right.$$
$$\left. - \frac{1}{\overline{p}_0}(c_3E\gamma + c_4\tau) - (c_1 + c_4)\sum_{n=1}^{\infty}(1-\overline{p}_1^n)\int_{n\tau}^{(n+1)\tau}\overline{F}(t)\,dt\right)\Bigg/\left(\frac{1}{\overline{p}_0}(E\gamma+\tau) + \right. \tag{2.89}$$
$$\left. + (E\gamma+\tau)\frac{\overline{p}_0 - p_1}{\overline{p}_0\overline{p}_1}\sum_{n=1}^{\infty}\overline{p}_1^n\overline{F}(n\tau) + E\beta\right),$$

$$C(\tau) = \left(c_2E\beta - c_4E\alpha + (c_4\tau + c_3E\gamma)\frac{\overline{p}_0 - p_1}{\overline{p}_0\overline{p}_1}\sum_{n=1}^{\infty}\overline{p}_1^n\overline{F}(n\tau) + \right.$$
$$\left. + c_4\sum_{n=1}^{\infty}(1-\overline{p}_1^n)\int_{n\tau}^{(n+1)\tau}\overline{F}(t)\,dt + \frac{1}{\overline{p}_0}(c_3E\gamma + c_4\tau)\right)\Bigg/$$
$$\left(E\alpha - \sum_{n=1}^{\infty}(1-\overline{p}_1^n)\int_{n\tau}^{(n+1)\tau}\overline{F}(t)\,dt\right). \tag{2.90}$$

Consider some specific cases.

Let $p_0 = 0, p_1 \neq 1$. Then formulas (2.80)–(2.82) take the form:

$$T_+ = \frac{E\alpha - \frac{p_1}{\bar{p}_1}\int_0^\infty \bar{F}(y)H_{\tilde{r}}(y)dy}{1+\int_0^\infty \bar{F}(z)dH_{\tilde{r}}(z)},$$

$$T_- = \frac{(E\delta + E\gamma)\left(1+\int_0^\infty \bar{F}(z)dH_{\tilde{r}}(z)\right) + E\beta - E\alpha + \frac{p_1}{\bar{p}_1}\int_0^\infty H_{\tilde{r}}(t)\bar{F}(t)dt}{1+\int_0^\infty \bar{F}(z)dH_{\tilde{r}}(z)},$$

$$K_a = \frac{E\alpha - \frac{p_1}{\bar{p}_1}\int_0^\infty \bar{F}(y)H_{\tilde{r}}(y)dy}{(E\delta + E\gamma)\left(1+\int_0^\infty \bar{F}(z)dH_{\tilde{r}}(z)\right) + E\beta}.$$

Let $p_1 = 0, p_0 \neq 1$, then $\tilde{r}(t) = r(t)$. Consequently, $H_{\tilde{r}}(t) = H_r(t)$ and (2.80)–(2.82) can be rewritten as follows:

$$T_+ = \frac{E\alpha}{1+\int_0^\infty \bar{F}(z)dH_r(z)}, T_- = \frac{(E\delta + E\gamma)\left(\frac{1}{\bar{p}_0}+\int_0^\infty \bar{F}(z)dH_r(z)\right) + E\beta - E\alpha}{1+\int_0^\infty \bar{F}(z)dH_r(z)},$$

$$K_a = \frac{E\alpha - \frac{p_1}{\bar{p}_1}\int_0^\infty \bar{F}(y)H_r(y)dy}{(E\delta + E\gamma)\left(\frac{1}{\bar{p}_0}+\int_0^\infty \bar{F}(z)dH_r(z)\right) + E\beta}.$$

Let $p_1 = 0; p_0 = 0$. In this case, formulas (2.80)–(2.82) transform into the following:

$$T_+ = \frac{E\alpha}{1+\int_0^\infty \bar{F}(z)dH_r(z)}, T_- = \frac{E\beta - E\alpha + (E\delta + E\gamma)\left(1+\int_0^\infty \bar{F}(z)dH_r(z)\right)}{1+\int_0^\infty \bar{F}(z)dH_r(z)},$$

$$K_a = \frac{E\alpha}{E\beta + (E\delta + E\gamma)\left(1+\int_0^\infty \bar{F}(z)dH_r(z)\right)}. \tag{2.91}$$

The expressions (2.91) coincide with (2.10)–(2.12).

TABLE 2.5 The Values of $K_a(\tau)$, $S(\tau)$, and $C(\tau)$ for the System With Unreliable Control

Initial data						Results			
Type of RV α distribution	$E\alpha$, h	$E\beta$, h	$E\gamma$, h	p_1	p_0	τ, h	$K_a(\tau)$	$S(\tau)$, c.u/h	$C(\tau)$, c.u/h
Exponential	60	0.5	0.2	0	0	4.833	0.916	1.739	0.101
Exponential	60	0.5	0.2	0.3	0.25	5.094	0.867	1.568	1.292
Erlangian of the second order	60	0.5	0.2	0.3	0.25	5.911	0.904	1.684	0.978
Erlangian of the second order	60	0.5	0.2	0.2	0.25	5.007	0.903	1.688	0.817

Example. Initial data and calculation results for stationary availability factor $K_a(\tau)$, average income $S(\tau)$ and average expenses $C(\tau)$, according to formulas (2.88)–(2.90), for the system studied are represented in Table 2.5.

The data $c_1 = 2$ c.u., $c_2 = 3$ c.u., $c_3 = 1$ c.u., and $c_4 = 1$ c.u. were taken to calculate average specific income and expenses.

2.5 THE SYSTEM MODEL WITH COMPONENT DEACTIVATION AND PREVENTIVE RESTORATION

2.5.1 System Description

The system S we consider consists of one component, performing certain functions, and of control equipment. The system operates as follows. At the initial moment, the component starts operating, the control is activated. The component TF is RV α with DF $F(t) = P\{\alpha \le t\}$ and DD $f(t)$. The component control is carried out in random time δ with RV $R(t) = P\{\delta \le t\}$ and DD $r(t)$. The component failure can be detected while control execution (latent failure). While control execution, the component is deactivated. The control duration is RV γ with DF $V(t) = P\{\gamma \le t\}$ and DD $v(t)$. If the component is treated as operable while control execution, and its total operating time has exceeded the given value $\mu > 0$, the preventive component restoration is carried out. In case total operating time is less than μ, the component operation goes on. If the component latent failure is detected while control, its ordinary restoration is carried out. Both ordinary and preventive control duration is RV β with DF

$G(t) = P\{\beta \le t\}$ and DD $g(t)$. The control is deactivated for the period of any kind of restoration. After restoration, all the component properties are completely restored. The RV $\alpha,\ \beta,\ \delta,\ \gamma$ are assumed to be independent and have finite expectations.

2.5.2 Semi-Markov Model Building

Let us describe the system operation by means of semi-Markov process $\xi(t)$ with the discrete–continuous phase state space. Let us introduce the set E of semi-Markov system states:

$$E = \{111, 211x, 2\hat{1}0x, 2\tilde{2}2, 101x, 2\hat{0}0, 222\}.$$

Let us write out the meaning of state codes:

111 – component begins operating, control is on;
$2\hat{1}0x$ – control has begun; the component is operable, deactivated, its total operation time from the previous restoration moment (regardless of control time) is $x > 0$;
$211x$ – control has ended, component is operable; its total operating time is less than μ, equals $x > 0$, the component continues operating;
$2\tilde{2}2$ – control has ended; the component is operable, its total operation time is not less than μ, the preventive component restoration has begun; the control is suspended;
$101x$ – the component fails; time $x > 0$ is left till the control beginning;
$2\hat{0}0$ – control has begun; failed component is deactivated;
222 – control has ended; latent failure is detected; ordinary restoration begins, control is suspended.

Note, in the states $2\hat{1}0x$, $211x$ «backward» time, whereas in $101x$ «forward» time is used.

Time diagram and system transition graph are represented in Figures 2.17 and 2.18, respectively.

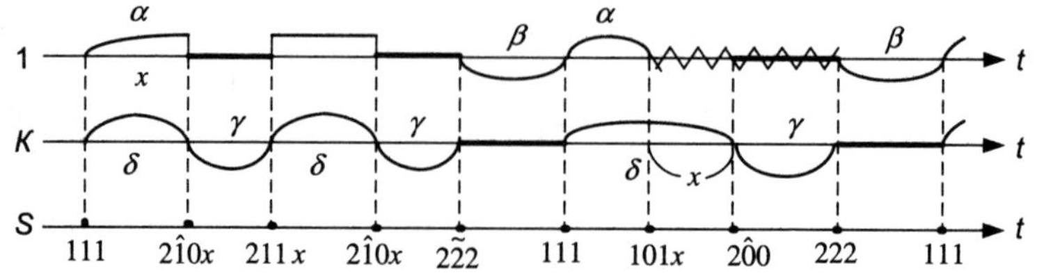

FIGURE 2.17 Time diagram of system operation

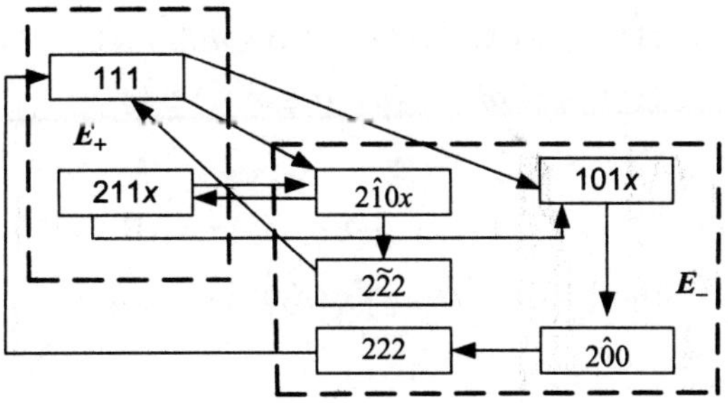

FIGURE 2.18 System transition graph

Introduce RV $[\alpha - x]^+$ with DF $F_x(t) = \dfrac{F(x+t) - F(x)}{\bar{F}(x)}$ and DD $\dfrac{f(x+t)}{\bar{F}(x)}$. In Ref. [3] this RV is called residual operating TF with age x.

Let us define system sojourn times in states. For instance, sojourn time θ_{211x} in $211x$ is determined by two factors: residual time $[\alpha - x]^+$ till the component latent failure and control periodicity δ. Consequently, $\theta_{211x} = \delta \wedge [\alpha - x]^+$. In the same way, system sojourn times in the rest of states are defined as follows:

$$\theta_{111} = \alpha \wedge \delta, \theta_{2\hat{1}0x} = \gamma, \theta_{2\tilde{2}2} = \beta, \theta_{101x} = x, \theta_{2\hat{0}0} = \gamma, \theta_{222} = \beta. \quad (2.92)$$

Let us describe system transition events. The events of transitions form the states 111, $211x$ are illustrated in Figures 2.19 and 2.20 and are defined by the expressions (2.93) and (2.94), respectively.

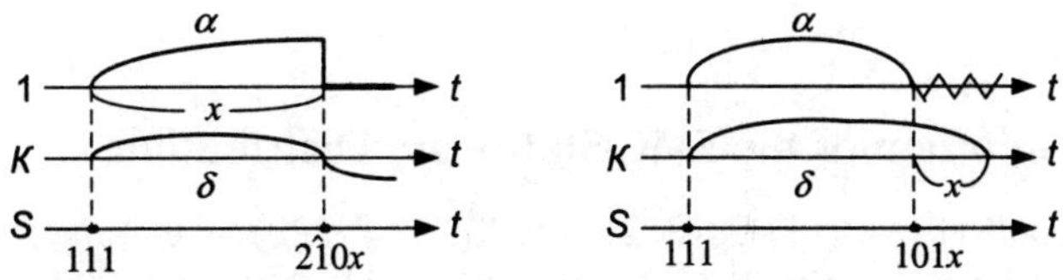

FIGURE 2.19 Transition events from the state 111

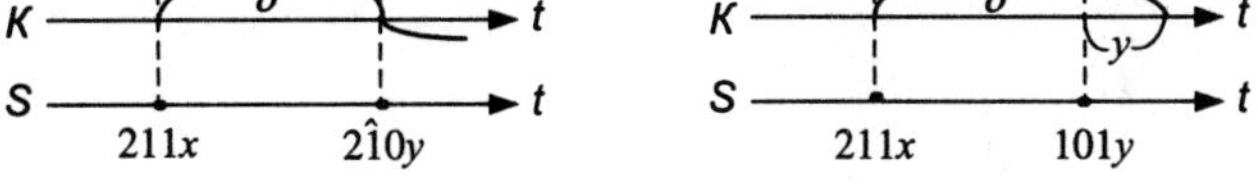

FIGURE 2.20 Transition events from the state $211x$

$$\begin{aligned}&\{111\to 2\hat{1}0dx\}=\{\alpha>\delta,\delta\in dx\},x>0;\\&\{111\to 101dx\}=\{\delta-\alpha\in dx\},x>0;\end{aligned} \tag{2.93}$$

$$\begin{aligned}&\left\{211x\to 2\hat{1}0dy\right\}=\left\{[\alpha-x]^{+}>\delta,\delta\in dy-x\right\},0<x<\mu,y>x;\\&\{211x\to 101dy\}=\left\{\delta-[\alpha-x]^{+}\in dy\right\},0<x<\mu,y>0.\end{aligned} \tag{2.94}$$

Transitions $2\hat{1}0x\to 211x$, $101x\to 2\hat{0}0$, $2\hat{0}0\to 222$, $222\to 111$ occur with the unity probability.

Formulas (2.93) and (2.94) allow to obtain following EMC $\{\xi_n;n\geq 0\}$ transition probabilities:

$$\begin{aligned}&p_{111}^{2\hat{1}0dx}=P\{\alpha>\delta,\delta\in dx\}=\bar{F}(x)r(x)dx,x>0;\\&p_{111}^{101dx}=P\{\delta-\alpha\in dx\}=\int_0^\infty r(x+t)f(t)dtdx,x>0;\\&p_{211x}^{2\hat{1}0dy}=P\left\{[\alpha-x]^{+}>\delta,\delta\in dy-x\right\}=\frac{\bar{F}(y)r(y-x)}{\bar{F}(x)}dy,0<x<\mu,y>x;\\&p_{211x}^{101dy}=P\left\{\delta-[\alpha-x]^{+}\in dy\right\}=\int_0^\infty\frac{f(x+t)}{\bar{F}(x)}r(y+t)dtdy,0<x<\mu,y>0;\\&P_{2\hat{1}0x}^{211x}=1,\text{if }0<x<\mu;\\&P_{2\hat{1}0x}^{2\tilde{2}2}=1,\text{if }x\geq\mu;\\&P_{2\tilde{2}2}^{111}=P_{101x}^{2\hat{0}0}=P_{2\hat{0}0}^{222}=P_{222}^{111}=1.\end{aligned} \tag{2.95}$$

2.5.3 Definition of the EMC Stationary Distribution

Let us denote by $\rho(111),\rho(2\tilde{2}2)$, $\rho(2\hat{0}0)$, $\rho(222)$ the values of EMC $\{\xi_n;n\geq 0\}$ stationary distribution in the states 111, $2\tilde{2}2$, $2\hat{0}0$, 222 and assume the existence of stationary densities $\rho(2\hat{1}0x)$, $\rho(211x)$, and $\rho(101x)$ for the states $2\hat{1}0x$, $211x$, and $101x$, respectively. Taking into account probabilities and probability densities of EMC $\{\xi_n;n\geq 0\}$ transitions (2.92), we can construct the system of integral equations to get the following stationary distribution:

$$\begin{cases} \rho(111) = \rho(2\tilde{2}2) + \rho(222), \\ \rho(2\hat{1}0x) = \rho(111)\bar{F}(x)r(x) + \int\limits_0^{x\wedge\mu} \frac{\bar{F}(x)r(x-y)}{\bar{F}(y)}\rho(211y)dy, x>0, \\ \rho(211x) = \rho(2\hat{1}0x), 0<x<\mu, \\ \rho(2\tilde{2}2) = \int\limits_\mu^\infty \rho(2\hat{1}0x)dx, \\ \rho(101x) = \rho(111)\int\limits_0^\infty r(x+y)f(y)dy + \\ \int\limits_0^\mu \rho(211y)dy\int\limits_0^\infty \frac{f(y+t)}{\bar{F}(y)}r(x+t)dt, x>0, \\ \rho(2\hat{0}0) = \int\limits_0^\infty \rho(101x)dx, \\ \rho(222) = \rho(2\hat{0}0), \\ 2\rho(111) + \rho(2\hat{0}0) + \int\limits_0^\infty \left(\rho(101x) + \rho(2\hat{1}0x)\right)dx + \int\limits_0^\mu \rho(211x)dx = 1. \end{cases} \tag{2.96}$$

The last equation in (2.96) is a normalization requirement.

Applying the method of successive approximations, one can prove the system of equations (2.96) has the following solution:

$$\begin{cases} \rho(111) = \rho_0, \\ \rho(2\hat{1}0x) = \rho(211x) = \rho_0 h_r(x)\bar{F}(x), 0<x<\mu, \\ \rho(2\hat{1}0x) = \rho_0\left(r(x) + \int\limits_0^\mu r(x-y)h_r(y)dy \right)\bar{F}(x), x\ge\mu, \\ \rho(2\tilde{2}2) = \rho_0\left(\int\limits_\mu^\infty \bar{F}(x)r(x)dx + \int\limits_\mu^\infty \bar{F}(x)dx\int\limits_0^\mu r(x-y)h_r(y)dy \right), \\ \rho(101x) = \rho_0\left(\int\limits_0^\infty r(x+y)f(y)dy + \int\limits_0^\mu h(y)dy\int\limits_0^\infty f(y+t)\, r(x+t)dt \right), \\ \rho(2\hat{0}0) = \rho(222) = \rho_0\left(\int\limits_0^\infty \bar{R}(x)f(x)dx + \int\limits_0^\infty \bar{R}(t)dt\int\limits_0^\mu f(y+t)h_r(y)dy \right), \end{cases} \tag{2.97}$$

where $h_r(t) = \sum\limits_{n=1}^\infty r^{*(n)}(t)$ is the density of renewal function $H_r(t) = \sum\limits_{n=1}^\infty R^{*(n)}(t)$ of the renewal process generated by RV δ; the constant ρ_0 is found from the normalization requirement.

2.5.4 Definition of the System Stationary Characteristics

Let us split the phase state space E into the following two subsets:

$E_+ = \{111, 211x\}$ – system up-state;

$E_- = \{2\hat{1}0x, 2\tilde{2}2, 101x, 2\hat{0}0, 222\}$ – system failure.

Average stationary operating TF T_+ and average stationary RT T_- can be determined by means of (1.17) and (1.18), respectively.

Applying formulas (2.92), we define average sojourn times in the following states:

$$m(111) = \int_0^\infty \bar{F}(1)\bar{R}(t)\,dt, m(211x) = \int_0^\infty \frac{\bar{F}(x+t)}{\bar{F}(x)}\bar{R}(t)\,dt, m(2\hat{1}0x) = E\gamma,$$
$$m(2\tilde{2}2) = E\beta, m(101x) = x, m(2\hat{0}0) = E\gamma, m(222) = E\beta. \tag{2.98}$$

With regard to (2.95), (2.97), and (2.98), we can obtain expressions from (1.17) and (1.18).

$$\int_{E_+} m(e)\rho(de) = m(111)\rho(111) + \int_0^\mu m(211x)\rho(211x)dx =$$
$$= \rho_0\left(\int_0^\infty \bar{F}(t)\bar{R}(t)\,dt + \int_0^\mu h_r(x)\,dx \int_0^\infty \bar{F}(x+t)\bar{R}(t)\,dt\right).$$

By changing the integration order we get the following:

$$\int_{E_+} m(e)\rho(de) = \rho_0 \int_0^\infty \bar{R}(t)\,dt \int_0^\mu \bar{F}(x+t)\,d\hat{H}_r(x), \tag{2.99}$$

where $\hat{H}_r(x) = \begin{cases} 1 + H_r(x), & x > 0, \\ 0, & x = 0. \end{cases}$

Next,

$$\int_{E_-} m(e)\rho(de) = m(2\tilde{2}2)\rho(2\tilde{2}2) + m(2\hat{0}0)\rho(2\hat{0}0) + m(222)\rho(222) +$$
$$+\int_0^\infty m(2\hat{1}0x)\rho(2\hat{1}0x)\,dx + \int_0^\infty m(101x)\rho(101x)\,dx =$$
$$= \rho_0\left(E\beta\left(\int_\mu^\infty \bar{F}(x)r(x)\,dx + \int_\mu^\infty \bar{F}(x)\,dx\int_0^\mu r(x-y)h_r(y)\,dy\right) + \right.$$
$$+E\gamma\left(\int_0^\infty \bar{R}(x)f(x)\,dx + \int_0^\infty \bar{R}(t)\,dt\int_0^\mu f(y+t)h_r(y)\,dy\right) +$$
$$+E\beta\left(\int_0^\infty \bar{R}(x)f(x)\,dx + \int_0^\infty \bar{R}(t)\,dt\int_0^\mu f(y+t)h_r(y)\,dy\right) +$$
$$+E\gamma\left(\int_0^\mu \bar{F}(x)h_r(x)\,dx + \int_\mu^\infty \bar{F}(x)r(x)\,dx + \int_\mu^\infty \bar{F}(x)\,dx\int_0^\mu r(x-y)h_r(y)\,dy\right) +$$
$$\left.+\int_0^\infty x\,dx\left(\int_0^\infty r(x+y)f(y)\,dy + \int_0^\mu h_r(y)\,dy\int_0^\infty f(y+t)r(x+t)\,dt\right)\right).$$

Simplifying the latter expression, we have:

$$\int_{E_-} m(e)\rho(de) = \rho_0\left(E\beta + (E\delta + E\gamma)\int_0^{\mu} \bar{F}(t)\,d\hat{H}_r(t) - \right. \\ \left. - \int_0^{\infty} \bar{R}(t)\,dt \int_0^{\mu} \bar{F}(t+x)\,d\hat{H}_r(x) \right). \tag{2.100}$$

Here, the formula (1.3) was applied.
Consequently,

$$\int_E m(e)\rho(de) = \rho_0\left(E\beta + (E\delta + E\gamma)\int_0^{\mu} \bar{F}(x)\,d\hat{H}_r(x) \right). \tag{2.101}$$

Let us define the denominator of (1.17) and (1.18).

$$\int_{E_+} P(e,E_-)\rho(de) = P(111,E_-)\rho(111) + \int_0^{\infty} P(211x,E_-)\rho(211x)\,dx = \\ = \rho_0\left(\int_0^{\infty} dx \int_0^{\infty} r(x+t)f(t)\,dt + \int_0^{\infty} \bar{F}(t)r(t)\,dt + \int_0^{\mu} h_r(x)\,dx \int_x^{\infty} \bar{F}(y)r(y-x)\,dy + \right. \\ \left. + \int_0^{\infty} dy \int_0^{\mu} h_r(x)\,dx \int_0^{\infty} r(y+t)f(t+x)\,dt \right) = \rho_0 \int_0^{\mu} \bar{F}(x)\,d\hat{H}_r(x). \tag{2.102}$$

By substituting the expressions (2.99)–(2.102) into (1.16)–(1.18), we get:

$$T_+ = \frac{\int_0^{\infty} \bar{R}(t)\,dt \int_0^{\mu} \bar{F}(x+t)\,d\hat{H}_r(x)}{\int_0^{\mu} \bar{F}(t)\,d\hat{H}_r(t)}, \tag{2.103}$$

$$T_- = \frac{E\beta + (E\delta + E\gamma)\int_0^{\mu} \bar{F}(t)\,d\hat{H}_r(t) - \int_0^{\infty} \bar{R}(t)\,dt \int_0^{\mu} \bar{F}(t+x)\,d\hat{H}_r(t)}{\int_0^{\mu} \bar{F}(t)\,d\hat{H}_r(t)}, \tag{2.104}$$

$$K_a = \frac{\int_0^{\infty} \bar{R}(t)\,dt \int_0^{\mu} \bar{F}(t+x)\,d\hat{H}_r(x)}{E\beta + (E\delta + E\gamma)\int_0^{\mu} \bar{F}(t)\,d\hat{H}_r(t)}. \tag{2.105}$$

Now we pass to the definition of system stationary efficiency characteristics: average specific income S per calendar time unit and average specific expenses C per time unit of system up-state. The functions $f_s(e), f_c(e)$ are as follows:

$$f_s(e)=\begin{cases} c_1, & e\in\{111,211x\},\\ -c_2, & e\in\{2\tilde{2}2,222\},\\ -c_3, & e\in\{2\hat{1}0x,2\hat{0}0\},\\ -c_4, & e=101x,\end{cases}\qquad f_c(e)=\begin{cases} 0, & e\in\{111,211x\},\\ c_2, & e\in\{2\tilde{2}2,222\},\\ c_3, & e\in\{2\hat{1}0x,2\hat{0}0\},\\ c_4, & e=101x.\end{cases}\tag{2.106}$$

Here, c_1 is the income per time unit of component up-state, c_2 are expenses per time unit of either preventive or ordinary component restoration, c_3 are expenses per time unit of control execution, and c_4 are expenses per time unit of component latent failure.

With regard to (2.95), (2.97), (2.98), and (2.106), average specific income per calendar time unit is

$$S=\frac{(c_1+c_4)\int_0^\infty \bar{R}(t)dt\int_0^\mu \bar{F}(x+t)d\hat{H}_r(x)-c_2E\beta-(c_3E\gamma+c_4E\delta)\int_0^\mu \bar{F}(t)d\hat{H}_r(t)}{E\beta+(E\delta+E\gamma)\int_0^\mu \bar{F}(t)d\hat{H}_r(t)}.\tag{2.107}$$

Average specific expenses per time unit of system up-state are

$$C=\frac{c_2E\beta+(c_4E\delta+c_3E\gamma)\int_0^\mu \bar{F}(t)d\hat{H}_r(t)-c_4\int_0^\infty \bar{R}(t)dt\int_0^\mu \bar{F}(x+t)d\hat{H}_r(x)}{\int_0^\infty \bar{R}(t)dt\int_0^\mu \bar{F}(x+t)d\hat{H}_r(x)}.\tag{2.108}$$

One should note that under $\mu\to+\infty$ the formulas (2.105), (2.107), and (2.108) transform into (2.12), (2.15), and (2.16), respectively.

Let us get formulas for reliability and efficiency characteristics of the given system in case of nonrandom control periodicity $\tau>0$. Then $R(t)=1(t-\tau)$, $E\delta=\tau$. The expressions (2.103)–(2.105), (2.107), and (2.108) will take the following form:

$$T_+=\frac{\sum_{n=0}^{[\mu/\tau]}\int_0^\tau \bar{F}(n\tau+t)dt}{\sum_{n=0}^{[\mu/\tau]}\bar{F}(n\tau)},\quad T_-=\frac{E\beta+(\tau+E\gamma)\sum_{n=0}^{[\mu/\tau]}\bar{F}(n\tau)-\sum_{n=0}^{[\mu/\tau]}\int_0^\tau \bar{F}(n\tau+t)dt}{\sum_{n=0}^{[\mu/\tau]}\bar{F}(n\tau)},$$

$$K_a=\frac{\sum_{n=0}^{[\mu/\tau]}\int_0^\tau \bar{F}(n\tau+t)dt}{E\beta+(\tau+E\gamma)\sum_{n=0}^{[\mu/\tau]}\bar{F}(n\tau)},\tag{2.109}$$

$$S=\frac{(c_1+c_4)\sum_{n=0}^{[\mu/\tau]}\int_0^{\tau}\overline{F}(n\tau+t)dt-c_2E\beta-(c_3E\gamma+c_4\tau)\sum_{n=0}^{[\mu/\tau]}\overline{F}(n\tau)}{E\beta+(\tau+E\gamma)\sum_{n=0}^{[\mu/\tau]}\overline{F}(n\tau)},\qquad(2.110)$$

$$C=\frac{c_2E\beta+(c_4\tau+c_3E\gamma)\sum_{n=0}^{[\mu/\tau]}\overline{F}(n\tau)-c_4\sum_{n=0}^{[\mu/\tau]}\int_0^{\tau}\overline{F}(n\tau+t)dt}{\sum_{n=0}^{[\mu/\tau]}\int_0^{\tau}\overline{F}(n\tau+t)dt},\qquad(2.111)$$

where $[x]$ is an integer part of x.

Example. To estimate the influence of preventive restoration on system characteristics, the initial data in the example coincide with the ones in Section 2.1.

Initial data and calculation results for $K_a(\tau,\mu)$, $S(\tau,\mu)$, and $C(\tau,\mu)$ by means of formulas (2.109), (2.110), and (2.111) are given in Table 2.6.

Comparative results that allow to estimate the impact of preventive restoration are given in Table 2.7. The initial data for the comparative calculations of $K_a(\tau,\mu)$ and $K_a(\tau)$ are: control periodicity $\tau = 5$ h, average TF $E\alpha = 70\,\text{h}$, average RT $E\beta = 0.50\,\text{h}$, and average control duration $E\gamma = 0.30\,\text{h}$.

TABLE 2.6 Values of $K_a(\tau,\mu)$, $S(\tau,\mu)$, and $C(\tau,\mu)$ Under $\tau = 5$ h

Initial data					Results		
Type of RV α distribution	$E\alpha$, h	$E\beta$, h	$E\gamma$, h	μ, h	$K_a(\tau,\mu)$	$S(\tau,\mu)$, c.u./h	$C(\tau,\mu)$, c.u./h
Exponential	60	0.083	0.10	10.5	0.935	6.436	0.117
Exponential	60	0.083	0.10	20.5	0.937	6.454	0.113
Erlangian of the 4th order	60	0.083	0.10	10.5	0.973	6.736	0.074
Erlangian of the 4th order	60	0.083	0.10	20.5	0.969	6.711	0.076
Erlangian of the 8th order	60	0.083	0.10	10.5	0.975	6.755	0.071
Erlangian of the 8th order	60	0.083	0.10	20.5	0.976	6.762	0.069
Veibull–Gnedenko	60	0.083	0.10	10.5	0.968	6.700	0.079
Veibull–Gnedenko	60	0.083	0.10	20.5	0.965	6.678	0.080

TABLE 2.7 Preventive Restoration Impact on Availability Factor

Type of RV α distribution	$K_a(\tau,\mu)$, calculated by (2.105)		$K_a(\tau)$, calculated by (2.17)
	μ, h	$K_a(\tau,\mu)$	
Exponential	20,5	0.886	0.898
Exponential	$\to\infty$	0.898	0.898
Erlangian of the 4th order	20.5	0.915	0.899
Erlangian of the 4th order	$\to\infty$	0.899	0.899
Veibull–Gnedenko	20.5	0.917	0.905
Veibull–Gnedenko	$\to\infty$	0.905	0.905

Chapter 3

Semi-Markov Models of Two-Component Systems with Regard to Control of Latent Failures

Chapter Outline

Semi-Markov Models. http://dx.doi.org/10.1016/B978-0-12-802212-2.00003-6
Copyright © 2015 Elsevier Inc. All rights reserved.

3.1 THE MODEL OF TWO-COMPONENT SERIAL SYSTEM WITH IMMEDIATE CONTROL AND RESTORATION

3.1.1 System Description

The system S, consisting of two serial (in reliability sense) components κ_1, κ_2 and of control unit, is considered. At the initial time moment, the components are operable, the control is on. Components time to failure (TF) are random variable (RV) α_1 and α_2 with distribution function (DF) $F_1(t) = P\{\alpha_1 \le t\}$, $F_2(t) = P\{\alpha_2 \le t\}$, and DD $f_1(t), f_2(t)$, respectively. The control is carried out in random time period δ with DF $R(t) = P\{\delta \le t\}$ and DD $r(t)$. The control of components operability is simultaneous. Failures are detected while control execution only (latent failures). Control and restoration are immediate, but after restoration, all the components properties get completely restored. RV α_1, α_2, δ are assumed to be independent and to have finite expectations.

3.1.2 Semi-Markov Model Building

To describe the system S operation let us introduce the following set E of system semi-Markov states:

$$E = \{3111, 3111x_1x_2, 1011x_2z, 2101x_1z, 1111x_2, 2111x_1, 1001z, 2001z\}.$$

The conceptual sense of codes is:

3111 – components κ_1, κ_2 begin to operate, control unit is on;
$3111x_1x_2$ – control is carried out, operable components κ_1, κ_2 continue to operate, times $x_1 > 0$, $x_2 > 0$ are left till their failures correspondingly;
$1011xz$ – component κ_1 has failed; component κ_2 is in up-state, time $x > 0$ is left till its failure, time $z > 0$ is left till the control beginning;
$2101xz$ – component κ_1 has failed, component κ_2 in up-state, time $x > 0$ is left till its failure, time $z > 0$ is left till the control beginning;
$1111x$ – component κ_1 has been restored and operates; component κ_2 continues to operate, time $x > 0$ is left till its failure, control unit is on;
$2111x$ – component κ_2 has been restored and operates; component κ_1 continues to operate, time $x > 0$ is left till its failure, control unit is on;
$1001z$ – component κ_1 has failed, κ_2 is in down-state, time $z > 0$ is left till the control beginning;
$2001z$ – component κ_2 has failed, κ_1 is in down-state, time $z > 0$ is left till the control beginning.

Time diagram of system operation and system transition graph are represented in Figures 3.1 and 3.2, respectively.

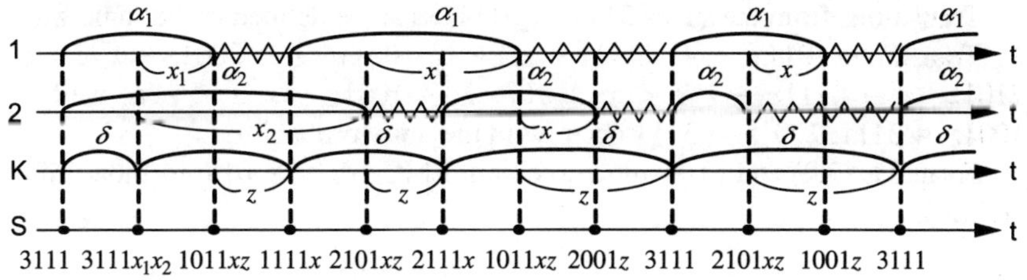

FIGURE 3.1 Time diagram of system operation

Let us define system sojourn times. For instance, sojourn time θ_{2111x} in the state $2111x$ is determined by three factors: residual time x_1 till component κ_1 latent failure, component κ_2 TF α_2, and control periodicity δ. Consequently, $\theta_{2111x} = x \wedge \alpha_2 \wedge \delta$. Similarly, sojourn times are defined for the rest of the states:

$$\theta_{3111} = \alpha_1 \wedge \alpha_2 \wedge \delta,\ \theta_{3111x_1x_2} = x_1 \wedge x_2 \wedge \delta,\ \theta_{1011xz} = x \wedge z,\ \theta_{2101xz} = x \wedge z,$$

$$\theta_{1111x} = \alpha_1 \wedge x \wedge \delta, \theta_{1001z} = z, \theta_{2001z} = z. \tag{3.1}$$

Let us describe system transition events. The events of transitions from the states 3111 and $2111x_1$ are determined by the relations (3.2) and (3.3) respectively:

$$\begin{aligned}
&\{3111 \to 3111dx_1dx_2\} = \{\alpha_1 - \delta \in dx_1, \alpha_2 - \delta \in dx_2\}, x_1 > 0, x_2 > 0;\\
&\{3111 \to 1011dxdz\} = \{\alpha_2 - \alpha_1 \in dx, \delta - \alpha_1 \in dz\}, x > 0, z > 0;\\
&\{3111 \to 2101dxdz\} = \{\alpha_1 - \alpha_2 \in dx, \delta - \alpha_2 \in dz\}, x > 0, z > 0;
\end{aligned} \tag{3.2}$$

$$\begin{aligned}
&\{2111x \to 3111dy_1dy_2\} = \{x - \delta \in dy_1, \alpha_2 - \delta \in dy_2\}, y_1 > 0, y_2 > 0;\\
&\{2111x \to 1011dydz\} = \{\alpha_2 - x \in dy, \delta - x \in dz\}, y > 0, z > 0;\\
&\{2111x \to 2101dydz\} = \{x - \alpha_2 \in dy, \delta - \alpha_2 \in dz\}, y > 0, z > 0.
\end{aligned} \tag{3.3}$$

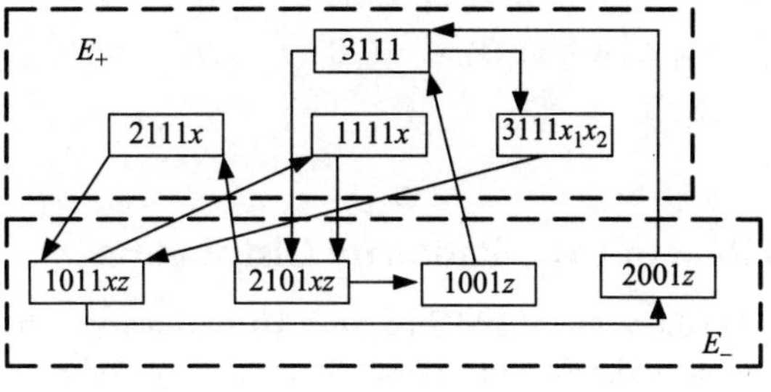

FIGURE 3.2 System transition graph for serial components connection

Transitions from the states $3111x_1x_2$, $1111x$ can be defined in the same way.

Transitions $1011x\ \ z \to 1111x - z$, $z < x$; $1011x\ \ z \to 2001z - x$, $x < z$; $2101x\ \ z \to 2111x - z$, $z < x$; $2101x\ \ z \to 1001z - x$, $x < z$, as well as $1001z \to 3111$, $2001z \to 3111$ occur with the unity probability.

Formulas (3.2) and (3.3) allow to obtain EMC $\{\xi_n; n \geq 0\}$ transition probabilities:

$$
\begin{aligned}
&p_{3111}^{3111x_1x_2} = \int_0^\infty f_1(x_1+t)f_2(x_2+t)r(t)dt, x_1>0, x_2>0;\\
&p_{3111}^{1011xz} = \int_0^\infty f_2(x+t)r(z+t)f_1(t)dt, x>0, z>0;\\
&p_{3111}^{2101xz} = \int_0^\infty f_1(x+t)r(z+t)f_2(t)dt, x>0, z>0;\\
&p_{3111x_1x_2}^{1011x_2-x_1z} = r(x_1+z), x_1<x_2, z>0;\\
&p_{3111x_1x_2}^{3111x_1-tx_2-t} = r(t), x_1<x_2, 0<t<x_1;\\
&p_{3111x_1x_2}^{2101x_2-x_1z} = r(x_2+z), x_2<x_1, z>0;\\
&p_{3111x_1x_2}^{3111x_1-tx_2-t} = r(t), x_2<x_1, 0<t<x_2;\\
&P_{1011xz}^{1111x-z} = 1, \text{if } z<x; p_{1011xz}^{2001z-x} = 1, \text{if } x<z;\\
&P_{2101xz}^{2111x-z} = 1, \text{if } z<x; p_{2101xz}^{1001z-x} = 1, \text{if } x<z;\\
&p_{1111x}^{2101yz} = f_1(y+x)r(z+x), y>0, z>0;\\
&p_{1111x}^{1011x-tz} = r(z+t)f_1(t), 0<t<x;\\
&p_{1111x}^{3111y_1x-t} = f_1(y_1+t)r(t), y_1>0, 0<t<x;\\
&p_{2111x}^{1011yz} = f_2(y+x)r(z+x), y>0, z>0;\\
&p_{2111x}^{2101x-tz} = r(z+t)f_2(t), 0<t<x;\\
&p_{2111x}^{3111x-ty_2} = f_2(y_2+t)r(t), y_2>0, 0<t<x; P_{1001z}^{3111} = P_{2001z}^{3111} = 1.
\end{aligned}
\tag{3.4}
$$

3.1.3 Definition of EMC Stationary Distribution

Denote by $\rho(3111)$ the value of EMC $\{\xi_n; n \geq 0\}$ stationary distribution in state 3111 and assume the existence of stationary densities $\rho(3111x_1x_2)$, $\rho(1011x\ z)$, $\rho(2101x\ z)$, $\rho(1111x)$, $\rho(2111x)$, $\rho(1001z)$, $\rho(2001z)$ for $3111x_1x_2$, $1011x\ z$, $2101x\ z$, $1111x$, $2111x$, $1001z$, $2001z$ correspondingly.

By using the EMC $\{\xi_n;\ n \geq 0\}$ probabilities and probability densities (3.4), taking into account (1.15), let us construct the system of integral equations to get the stationary distribution:

$$\begin{cases} \rho_0 = \rho(3111) = \int_0^\infty \rho(1001z)dz + \int_0^\infty \rho(2001z)dz, \\ \rho(3111x_1x_2) = \int_0^\infty \rho(3111x_1 + t\ x_2 + t)r(t)dt + \\ +\rho(3111)\int_0^\infty f_1(x_1+t)f_2(x_2+t)r(t)dt + \\ +\int_0^\infty \rho(1111x_2+t)f_1(x_1+t)r(t)dt + \int_0^\infty \rho(2111x_1+t)f_2(x_2+t)r(t)dt, \\ \rho(1011xz) = \rho(3111)\int_0^\infty f_1(t)f_2(x+t)r(z+t)dt + \\ +\int_0^\infty \rho(1111x+t)f_1(t)r(z+t)dt + \int_0^\infty \rho(2111t)f_2(x+t)r(z+t)dt + \\ +\int_0^\infty \rho(3111tx+t)r(z+t)dt, \\ \rho(2101xz) = \rho(3111)\int_0^\infty f_1(x+t)f_2(t)r(z+t)dt + \\ +\int_0^\infty \rho(2111x+t)f_2(t)r(z+t)dt + \int_0^\infty \rho(1111t)f_1(x+t)r(z+t)dt + \\ +\int_0^\infty \rho(3111x+tt)r(z+t)dt, \\ \rho(1111x) = \int_0^\infty \rho(1011x+tt)dt, \rho(2111x) = \int_0^\infty \rho(2101x+tt)dt, \\ \rho(1001z) = \int_0^\infty \rho(2101tz+t)dt, \rho(2001z) = \int_0^\infty \rho(1011tz+t)dt, \\ \rho_0 + \int_0^\infty\int_0^\infty \rho(3111x_1x_2)dx_1\,dx_2 + \int_0^\infty\int_0^\infty \rho(1011xz)dx\,dz + \int_0^\infty\int_0^\infty \rho(2101xz)dx\,dz + \\ +\int_0^\infty \rho(1111x)dx + \int_0^\infty \rho(2111x)dx + \int_0^\infty \rho(1001z)dz + \int_0^\infty \rho(2001z)dz = 1. \end{cases} \quad (3.5)$$

The last equation in the system (3.5) is a normalization requirement and the introduced notations are as follows:

$$\begin{aligned} &\rho_0 = \rho(3111), \varphi_1(x_1, x_2) = \rho(3111x_1x_2), \\ &\varphi_2(x,z) = \rho(1011x\,z), \varphi_3(x,z) = \rho(2101x\,z), \varphi_4(x) = \rho(1111x), \\ &\varphi_5(x) = \rho(2111x), \varphi_6(z) = \rho(1001z), \varphi_7(z) = \rho(2001z). \end{aligned} \quad (3.6)$$

Then the system (3.5) takes the following form:

$$\begin{cases} \rho_0 = \int_0^\infty \varphi_6(z)\,dz + \int_0^\infty \varphi_7(z)\,dz, \\ \varphi_1(x_1,x_2) = \rho_0 \int_0^\infty f_1(x_1+t) f_2(x_2+t) r(t)\,dt + \int_0^\infty \varphi_1(x_1+t, x_2+t) r(t)\,dt + \\ + \int_0^\infty \varphi_4(x_2+t) f_1(x_1+t) r(t)\,dt + \int_0^\infty \varphi_5(x_1+t) f_2(x_2+t) r(t)\,dt, \\ \varphi_2(x,z) = \rho_0 \int_0^\infty f_1(t) f_2(x+t) r(z+t)\,dt + \int_0^\infty \varphi_4(x+t) f_1(t) r(z+t)\,dt + \\ + \int_0^\infty \varphi_5(t) f_2(x+t) r(z+t)\,dt + \int_0^\infty \varphi_1(t, x+t) r(z+t)\,dt, \\ \varphi_3(x,z) = \rho_0 \int_0^\infty f_1(x+t) f_2(t) r(z+t)\,dt + \int_0^\infty \varphi_5(x+t) f_2(t) r(z+t)\,dt + \\ + \int_0^\infty \varphi_4(t) f_1(x+t) r(z+t)\,dt + \int_0^\infty \varphi_1(x+t, t) r(z+t)\,dt, \\ \varphi_4(x) = \int_0^\infty \varphi_2(x+t, t)\,dt, \\ \varphi_5(x) = \int_0^\infty \varphi_3(x+t, t)\,dt, \\ \varphi_6(z) = \int_0^\infty \varphi_3(t, z+t)\,dt, \\ \varphi_7(z) = \int_0^\infty \varphi_2(t, z+t)\,dt, \\ \rho_0 + \int_0^\infty\int_0^\infty \varphi_1(x_1,x_2)\,dx_1\,dx_2 + \int_0^\infty\int_0^\infty \varphi_2(x,z)\,dx\,dz + \int_0^\infty\int_0^\infty \varphi_3(x,z)\,dx\,dz + \\ + \int_0^\infty \varphi_4(x)\,dx + \int_0^\infty \varphi_5(x)\,dx + \int_0^\infty \varphi_6(z)\,dz + \int_0^\infty \varphi_7(z)\,dz = 1. \end{cases} \tag{3.7}$$

Let us indicate

$$\bar{i} = \begin{cases} 1, & if \quad i=2, \\ 2, & if \quad i=1. \end{cases} \tag{3.8}$$

To write down the solution of the system of equations (3.7) let us consider the following functions:

$h_r(t) = \sum_{n=1}^{\infty} r^{*(n)}(t)$ is the density of renewal function $H_r(t)$ of the renewal process generated by RV δ;

$v_r(z,x)=r(z+x)+\int_0^z r(z+x-s)h_r(s)\,ds$ is the distribution density of direct residual time for the renewal process generated by RV δ;

$h_i(t)=\sum_{n=1}^{\infty}\tilde{\gamma}_i^{*(n)}(t), i=1,2$ are densities of the renewal functions of renewal processes, generated by RV with densities $\tilde{\gamma}_i(t)=\int_0^t f_i(t)v_r(y,t-y)\,dy$;

$$\gamma_i(x,t)=\int_0^{\infty} f_i(x+z+t)v_r(t,z)\,dz+\int_0^{\infty} h_{\bar{i}}(y)\,dy\int_0^{\infty} f_i(x+z+y+t)v_r(t,z)\,dz, 1,2;$$

$$\pi_i(x,y)=\sum_{n=1}^{\infty} k_i^{*(n)}(x,y), 1=1,2,$$

where $k_i^{(1)}(x,y)=k_i(x,y)=\int_0^{\infty}\gamma_i(x,t)\gamma_{\bar{i}}(t,y)\,dt, k_i^{(n)}(x,y)=\int_0^{\infty}k_i(x,t)k_i^{(n-1)}(t,y)\,dt.$

Note the functions $\pi_i(x, y)$ are resolvents, appearing in the process of solution of Fredholm equations of the second kind (3.7).

One can prove (Appendix C) the system of equations (3.7) to have the following solution:

$$\varphi_4(x)=\rho_0\psi_2(x)=\rho_0\left(\int_0^{\infty} f_2(x+y)h_1(y)dy+\int_0^{\infty}\gamma_2(x,t)dt\int_0^{\infty} f_1(t+y)h_2(y)dy+\right.$$
$$\left.+\int_0^{\infty}\pi_2(x,y)dy\int_0^{\infty} f_2(y+t)h_1(t)dt+\int_0^{\infty}\pi_2(x,y)dy\int_0^{\infty}\gamma_2(y,t)dt\int_0^{\infty} f_1(t+z)h_2(z)dz\right),$$

$$\varphi_5(x)=\rho_0\psi_1(x)=\rho_0\left(\int_0^{\infty} f_1(x+y)h_2(y)dy+\int_0^{\infty}\gamma_1(x,t)dt\int_0^{\infty} f_2(t+y)h_1(y)dy+\right.$$
$$\left.+\int_0^{\infty}\pi_1(x,y)dy\int_0^{\infty} f_1(y+t)h_2(t)dt+\int_0^{\infty}\pi_1(x,y)dy\int_0^{\infty}\gamma_1(y,t)dt\int_0^{\infty} f_2(t+z)h_1(z)dz\right),$$

where $\psi_1(x)$, $\psi_2(x)$ – the functions are given in parenthesis,

$$\varphi_1(x_1,x_2)=\rho_0\left(\int_0^{\infty} f_1(x_1+y)f_2(x_2+y)h_r(y)dy+\right.$$
$$\left.+\int_0^{\infty}\psi_1(x_1+t)f_2(x_2+t)h_r(t)dt+\int_0^{\infty}\psi_2(x_2+t)f_1(x_1+t)h_r(t)dt\right),$$

$$\varphi_2(x,z)=\rho_0\left(\int_0^{\infty} f_1(y)f_2(x+y)v_r(y,z)dy+\right.$$
$$\left.+\int_0^{\infty}\psi_1(t)f_2(x+t)v_r(t,z)dt+\int_0^{\infty}\psi_2(x+t)f_1(t)v_r(t,z)dt\right),$$

$$\varphi_3(x,z)=\rho_0\left(\int_0^\infty f_1(x+y)f_2(y)v_r(y,z)dy+\right.$$
$$\left.+\int_0^\infty \psi_2(t)f_1(x+t)v_r(t,z)dt+\int_0^\infty \psi_1(x+t)f_2(t)v_r(t,z)dt\right),$$
$$\varphi_7(z)=\rho_0\left(\int_0^\infty dt\int_0^\infty f_1(y)f_2(t+y)v_g(y,t+z)dy+\right.$$
$$\left.+\int_0^\infty dt\int_0^\infty \psi_1(y)f_2(t+y)v_r(y,t+z)dy+\int_0^\infty dt\int_0^\infty \psi_2(t+y)f_1(y)v_r(y,t+z)dy\right), \tag{3.9}$$
$$\varphi_6(z)=\rho_0\left(\int_0^\infty dt\int_0^\infty f_1(t+y)f_2(y)v_r(y,t+z)dy+\right.$$
$$\left.+\int_0^\infty dt\int_0^\infty \psi_1(t+y)f_2(y)v_r(y,t+z)dy+\int_0^\infty dt\int_0^\infty \psi_2(y)f_1(t+y)v_r(y,t+z)dy\right).$$

The constant ρ_0 is found by means of normalization requirement.

3.1.4 Stationary Characteristics Definition

Let us split the phase state space E into the following two subsets:

$E_+ = \{3111, 3111x_1x_2, 1111x_2, 2111x_1\}$ – the system is in up-state;

$E_- = \{E = \{1011x_2z, 2101x_1z, 1001z, 2001z\}$ – the system is in down-state.

Applying (3.1), let us define system average sojourn times in states:

$$m(3111)=\int_0^\infty \bar F_1(t)\bar F_2(t)\bar R(t)dt, m(3111x_1x_2)=\int_0^{x_1\wedge x_2}\bar R(t)dt, m(1011xz)=x\wedge z,$$
$$m(2101xz)=x\wedge z, m(1111x)=\int_0^x \bar F_1(t)\bar R(t)dt, m(2111x)=\int_0^x \bar F_2(t)\bar R(t)dt,$$
$$m(1001z)=z, m(2001z)=z. \tag{3.10}$$

Average stationary operation TF T_+ and average stationary restoration time T_- can be obtained by formulas given in Chapter 1 (1.17 and 1.18).

With regard to (3.4), (3.9), and (3.10) we get the expressions from Chapter 1 (1.17 and 1.18).

$$\int_{E+} m(e)\rho(de)=m(3111)\rho(3111)+\int_0^\infty m(2111x)\rho(2111x)dx+$$
$$+\int_0^\infty m(1111x)\rho(1111x)dx+\int_0^\infty dx_1\int_0^\infty m(3111x_1x_2)\rho(3111x_1x_2)dx_2=$$

$$
\begin{aligned}
&= \rho_0 \int_0^\infty \bar{F}_1(t)\bar{F}_2(t)\bar{R}(t)dt + \rho_0 \int_0^\infty \psi_1(x)\,dx \int_0^x \bar{F}_2(t)\bar{R}(t)\,dt + \\
&+\rho_0 \int_0^\infty \psi_2(x)\,dx \int_0^x \bar{F}_1(t)\bar{R}(t)\,dt + \int_0^\infty dx_1 \int_0^{x_1} \varphi_3(x_1,x_2)\,dx_2 \int_0^{x_2} \bar{R}(t)\,dt + \\
&+\int_0^\infty dx_1 \int_{x_1}^\infty \varphi_3(x_1,x_2)\,dx_2 \int_0^{x_1} \bar{R}(t)\,dt = \rho_0 \Bigg(\int_0^\infty \bar{F}_1(t)\bar{F}_2(t)\bar{R}(t)dt + \int_0^\infty \psi_1(x)\,dx \int_0^x \bar{F}_2(t)\bar{R}(t)\,dt + \\
&+\int_0^\infty \psi_2(x)\,dx \int_0^x \bar{F}_1(t)\bar{R}(t)\,dt + \int_0^\infty dx_1 \int_0^{x_1} dx_2 \int_0^{x_2} \bar{R}(t)\,dt \int_0^\infty f_1(x_1+y)f_2(x_2+y)h_r(y)\,dy + \\
&+\int_0^\infty dx_1 \int_0^{x_1} dx_2 \int_0^{x_2} \bar{R}(t)\,dt \int_0^\infty \psi_1(x_1+y)f_2(x_2+y)h_r(y)\,dy + \\
&+\int_0^\infty dx_1 \int_0^{x_1} dx_2 \int_0^{x_2} \bar{R}(t)\,dt \int_0^\infty \psi_2(x_2+y)f_1(x_1+y)h_r(y)\,dy + \\
&+\int_0^\infty dx_1 \int_{x_1}^\infty dx_2 \int_0^{x_1} \bar{R}(t)\,dt \int_0^\infty f_1(x_1+y)f_2(x_2+y)h_r(y)dy + \\
&+\int_0^\infty dx_1 \int_{x_1}^\infty dx_2 \int_0^{x_1} \bar{R}(t)\,dt \int_0^\infty \psi_1(x_1+y)f_2(x_2+y)h_r(y)\,dy + \\
&+\int_0^\infty dx_1 \int_{x_1}^\infty dx_2 \int_0^{x_1} \bar{R}(t)\,dt \int_0^\infty \psi_2(x_2+y)f_1(x_1+y)h_r(y)\,dy \Bigg).
\end{aligned}
\tag{3.11}
$$

After some transformations, the expression (3.11) takes the following form:

$$
\int_{E_+} m(e)\rho(de) = \rho_0 \left(E(\alpha_1 \wedge \alpha_2) + \int_0^\infty \bar{F}_2(t)\bar{\Phi}_1(t)\,dt + \int_0^\infty \bar{F}_1(t)\bar{\Phi}_2(t)\,dt \right). \tag{3.12}
$$

Here, using the notation (3.8),

$$
\begin{aligned}
\bar{\Phi}_i(t) &= \int_0^\infty \bar{F}_i(t+y)h_{\bar{i}}(y)\,dy + \int_0^\infty \bar{\Gamma}_i(t,y)\,dy \int_0^\infty f_i(y+z)h_i(z)\,dz + \\
&+\int_0^\infty \bar{\Pi}_i(t,y)\,dy \int_0^\infty f_i(y+z)h_{\bar{i}}(z)\,dz + \int_0^\infty \bar{\Pi}_i(t,y)\,dy \int_0^\infty \gamma_i(y,z)\,dz \int_0^\infty f_{\bar{i}}(z+s)h_i(s)\,ds, i=1,2,
\end{aligned}
$$

where

$$
\bar{\Gamma}_i(t,y) = \int_t^\infty \gamma_i(x,y)\,dx, \bar{\Pi}_i(t,y) = \int_t^\infty \pi_i(x,y)\,dx, i=1,2.
$$

Next,

$$\int_{E_-} m(e)\rho(de) = \int_0^\infty x\,dx \int_x^\infty \rho(1011xz)\,dz + \int_0^\infty dx \int_0^x z\rho(1011xz)\,dz +$$

$$+\int_0^\infty x\,dx \int_x^\infty \rho(2101xz)\,dz + \int_0^\infty dx \int_0^x z\rho(2101xz)\,dz + \int_0^\infty z\rho(2001z)\,dz + \int_0^\infty z\rho(1001z)\,dz =$$

$$= \int_0^\infty x\,dx \int_x^\infty \varphi_2(x,z)\,dz + \int_0^\infty dx \int_0^x z\varphi_2(x,z)\,dz + \int_0^\infty x\,dx \int_x^\infty \varphi_3(x,z)\,dz + \int_0^\infty dx \int_0^x z\varphi_3(x,z)\,dz +$$

$$+\int_0^\infty z\varphi_7(z)\,dz + \int_0^\infty z\varphi_6(z)\,dz = \rho_0 \left(\int_0^\infty x\,dx \int_x^\infty dz \int_0^\infty f_1(y) f_2(x+y) v_r(y,z)\,dy + \right.$$

$$+\int_0^\infty x\,dx \int_x^\infty dz \int_0^\infty \psi_1(y) f_2(x+y) v_r(y,z)\,dy + \int_0^\infty x\,dx \int_x^\infty dz \int_0^\infty \psi_2(x+y) f_1(y) v_r(y,z)\,dy +$$

$$+\int_0^\infty dx \int_0^x z\,dz \int_0^\infty f_1(y) f_2(x+y) v_r(y,z)\,dy + \int_0^\infty dx \int_0^x z\,dz \int_0^\infty \psi_1(y) f_2(x+y) v_r(y,z)\,dy +$$

$$+\int_0^\infty dx \int_0^x z\,dz \int_0^\infty \psi_2(x+y) f_1(y) v_r(y,z)\,dy + \int_0^\infty x\,dx \int_x^\infty dz \int_0^\infty f_1(x+y) f_2(y) v_r(y,z)\,dy +$$

$$+\int_0^\infty x\,dx \int_x^\infty dz \int_0^\infty \psi_1(x+y) f_2(y) v_r(y,z)\,dy + \int_0^\infty x\,dx \int_x^\infty dz \int_0^\infty \psi_2(y) f_1(x+y) v_r(y,z)\,dy +$$

$$+\int_0^\infty dx \int_0^x z\,dz \int_0^\infty f_1(x+y) f_2(y) v_r(y,z)\,dy + \int_0^\infty dx \int_0^x z\,dz \int_0^\infty \psi_1(x+y) f_2(y) v_r(y,z)\,dy +$$

$$+\int_0^\infty dx \int_0^x z\,dz \int_0^\infty \psi_2(y) f_1(x+y) v_r(y,z)\,dy + \int_0^\infty z\,dz \int_0^\infty dx \int_0^\infty f_1(y) f_2(x+y) v_r(y,x+z)\,dy +$$

$$+\int_0^\infty z\,dz \int_0^\infty dx \int_0^\infty \psi_1(y) f_2(x+y) v_r(y,x+z)\,dy + \int_0^\infty z\,dz \int_0^\infty dx \int_0^\infty \psi_2(x+y) f_1(y) v_r(y,x+z)\,dy +$$

$$+\int_0^\infty z\,dz \int_0^\infty dx \int_0^\infty f_1(x+y) f_2(y) v_r(y,x+z)\,dy + \int_0^\infty z\,dz \int_0^\infty dx \int_0^\infty \psi_1(x+y) f_2(y) v_r(y,x+z)\,dy +$$

$$\left. +\int_0^\infty z\,dz \int_0^\infty dx \int_0^\infty \psi_2(y) f_1(x+y) v_r(y,x+z)\,dy \right). \tag{3.13}$$

After simplifying, the expression (3.13) can be rewritten as follows:

$$\int_{E_-} m(e)\rho(de) = \rho_0 \left(E\delta \int_0^\infty \bar{F}_1(y)\bar{F}_2(y)\,d\hat{H}_r(y) + E\delta \int_0^\infty \bar{\Phi}_1(y)\bar{F}_2(y)\,d\hat{H}_r(y) + \right.$$

$$\left. +E\delta \int_0^\infty \bar{\Phi}_2(y)\bar{F}_1(y)\,d\hat{H}_r(y) - E(\alpha_1 \wedge \alpha_2) - \int_0^\infty \bar{\Phi}_1(y)\bar{F}_2(y)\,dy - \int_0^\infty \bar{\Phi}_2(y)\bar{F}_1(y)\,dy \right). \tag{3.14}$$

Then,

$$\int_E m(e)\rho(de) = \rho_0 E\delta \left(\int_0^\infty \bar{F}_1(y)\bar{F}_2(y)\,d\hat{H}_r(y) + \int_0^\infty \bar{\Phi}_1(y)\bar{F}_2(y)\,d\hat{H}_r(y) + \right.$$

$$\left. +\int_0^\infty \bar{\Phi}_2(y)\bar{F}_1(y)\,d\hat{H}_r(y) \right),$$

where $\hat{H}_r(t)$ is the function given by Chapter 1 (1.4): $\hat{H}_r(t)=\begin{cases}1+H_r(t), & t>0,\\ 0, t=0.\end{cases}$

Next,

$$\begin{aligned}
&\int_{E_+} P(e,E_-)\rho(de)=\rho_0\int_0^\infty dx\int_0^\infty dz\int_0^\infty f_2(t)f_1(x+t)r(z+t)dt+\\
&+\rho_0\int_0^\infty dx\int_0^\infty dz\int_0^\infty f_1(t)f_2(x+t)r(z+t)dt+\int_0^\infty dx_1\int_{x_1}^\infty \rho(3111x_1x_2)dx_2\int_0^\infty r(x_1+z)dz+\\
&+\int_0^\infty dx_1\int_0^{x_1}\rho(3111x_1x_2)dx_2\int_0^\infty r(x_2+z)dz+\int_0^\infty \rho(2111x)dx\int_0^\infty f_2(x+y)dy\int_0^\infty r(x+z)dz+\\
&+\int_0^\infty \rho(2111x)dx\int_0^x f_2(x-y)dy\int_0^\infty r(x-y+z)dz+\int_0^\infty \rho(1111x)dx\int_0^\infty f_1(x+y)dy\int_0^\infty r(x+z)dz+\\
&+\int_0^\infty \rho(1111x)dx\int_0^x f_1(x-y)dy\int_0^\infty r(x-y+z)dz=\\
&=\rho_0\left(\int_0^\infty dx\int_0^\infty dz\int_0^\infty f_2(t)f_1(x+t)r(z+t)dt+\int_0^\infty dx\int_0^\infty dz\int_0^\infty f_1(t)f_2(x+t)r(z+t)dt+\right.\\
&+\int_0^\infty dx_1\int_0^{x_1}\bar{R}(x_2)dx_2\int_0^\infty f_1(x_1+y)f_2(x_2+y)h_r(y)dy+\\
&+\int_0^\infty dx_1\int_0^{x_1}\bar{R}(x_2)dx_2\int_0^\infty \psi_1(x_1+t)f_2(x_2+t)h_r(t)dt+\\
&+\int_0^\infty dx_1\int_0^{x_1}\bar{R}(x_2)dx_2\int_0^\infty \psi_2(x_2+t)f_1(x_1+t)h_r(t)dt+\\
&+\int_0^\infty dx_1\int_{x_1}^\infty \bar{R}(x_1)dx_2\int_0^\infty f_1(x_1+y)f_2(x_2+y)h_r(y)dy+\\
&+\int_0^\infty dx_1\int_{x_1}^\infty \bar{R}(x_1)dx_2\int_0^\infty \psi_1(x_1+t)f_2(x_2+t)h_r(t)dt+\\
&+\int_0^\infty dx_1\int_{x_1}^\infty \bar{R}(x_1)dx_2\int_0^\infty \psi_2(x_2+t)f_1(x_1+t)h_r(t)dt+\\
&+\int_0^\infty \psi_1(x)dx\int_0^\infty f_2(x+y)dy\int_0^\infty r(x+z)dz+\int_0^\infty \psi_1(x)dx\int_0^x f_2(x-y)dy\int_0^\infty r(x-y+z)dz+\\
&\left.+\int_0^\infty \psi_2(x)dx\int_0^\infty f_1(x+y)dy\int_0^\infty r(x+z)dz+\int_0^\infty \psi_2(x)dx\int_0^x f_1(x-y)dy\int_0^\infty r(x-y+z)dz\right).
\end{aligned} \tag{3.15}$$

As a result of manipulations, (3.15) takes the following form:

$$\int_{E_+} P(x,E_-)\rho(dx)=\rho_0\left(1+\bar{\Phi}_1(0)+\bar{\Phi}_2(0)\right). \tag{3.16}$$

Here, with regard to the definition of the functions $\bar{\Phi}_i(t)$,

$$\overline{\Phi}_i(0)=\int_0^\infty \overline{F}_i(y)h_{\bar{i}}(y)\,dy+\int_0^\infty \overline{\Gamma}_i(0,y)\,dy\int_0^\infty f_{\bar{i}}(y+z)h_i(z)\,dz+$$
$$+\int_0^\infty \overline{\Pi}_i(0,y)\,dy\int_0^\infty f_i(y+z)h_{\bar{i}}(z)\,dz+\int_0^\infty \overline{\Pi}_i(0,y)\,dy\int_0^\infty \gamma_i(y,z)\,dz\int_0^\infty f_{\bar{i}}(z+s)h_i(s)\,ds,$$

where

$$\overline{\Gamma}_i(0,y)=\int_0^\infty \gamma_i(x,y)\,dx,\overline{\Pi}_i(0,y)=\int_0^\infty \pi_i(x,y)\,dx,i=1,2.$$

Thus, with regard to (3.12), (3.14), and (3.16), average stationary operating TF T_+ is:

$$T_+=\frac{E(\alpha_1\wedge\alpha_2)+\int_0^\infty \overline{F}_2(t)\overline{\Phi}_1(t)\,dt+\int_0^\infty \overline{F}_1(t)\overline{\Phi}_2(t)\,dt}{1+\overline{\Phi}_1(0)+\overline{\Phi}_2(0)}, \tag{3.17}$$

average stationary restoration time T_- is as follows:

$$T_-=\left(E\delta\int_0^\infty \overline{F}_1(y)\overline{F}_2(y)\,d\hat{H}_r(y)+E\delta\int_0^\infty \overline{\Phi}_1(y)\overline{F}_2(y)\,d\hat{H}_r(y)+E\delta\int_0^\infty \overline{\Phi}_2(y)\overline{F}_1(y)\,d\hat{H}_r(y)-\right.$$
$$\left.-E(\alpha_1\wedge\alpha_2)-\int_0^\infty \overline{\Phi}_1(y)\overline{F}_2(y)\,dy-\int_0^\infty \overline{\Phi}_2(y)\overline{F}_1(y)\,dy\right)\Big/\left(1+\overline{\Phi}_1(0)+\overline{\Phi}_2(0)\right). \tag{3.18}$$

By means of formulas (3.17) and (3.18), stationary availability factor is defined in the following way:

$$K_a=\frac{E(\alpha_1\wedge\alpha_2)+\int_0^\infty \overline{F}_2(t)\overline{\Phi}_1(t)\,dt+\int_0^\infty \overline{F}_1(t)\overline{\Phi}_2(t)\,dt}{E\delta\left(\int_0^\infty \overline{F}_1(y)\overline{F}_2(y)\,d\hat{H}_r(y)+\int_0^\infty \overline{\Phi}_1(y)\overline{F}_2(y)\,d\hat{H}_r(y)+\int_0^\infty \overline{\Phi}_2(y)\overline{F}_1(y)\,d\hat{H}_r(y)\right)}. \tag{3.19}$$

Let us find system efficiency characteristics: average specific income S per calendar time unit, average specific expenses C per time unit of up-state. To do it, we apply the formulas given in Chapter 1 (1.22 and 1.23).

Let c_1 be the income per time unit of system up-state; c_2 be expenses for system failure per time unit.

For the given system, the functions $f_s(e), f_c(e)$ look like:

$$f_s(e)=\begin{cases} c_2, & e\in\{1011,\quad 3111x_1x_2\quad 1111x,\quad 2111x\},\\ -c_2, & e\in\{1011xz,\quad 2101xz,\quad 1001z,\quad 2001z\},\end{cases}$$

$$f_c(e)=\begin{cases}0, & e\in\{3111,\quad 3111x_1x_2,\quad 1111x,\quad 2111x\},\\ c_2, & e\in\{1011xz,\quad 2101xz,\quad 1001z,\quad 2001z\}.\end{cases}$$

Average specific income is given by the expression:

$$S=\left((c_1+c_2)\left(E(\alpha_1\wedge\alpha_2)+\int_0^\infty \bar F_2(t)\bar\Phi_1(t)\,dt+\int_0^\infty \bar F_1(t)\bar\Phi_2(t)\,dt\right)-\right.$$
$$-c_2E\delta\left(\int_0^\infty \bar F_1(y)\bar F_2(y)\,d\hat H_r(y)+\int_0^\infty \bar\Phi_1(y)\bar F_2(y)\,d\hat H_r(y)+\right.$$
$$\left.\left.\int_0^\infty \bar\Phi_2(y)\bar F_1(y)\,d\hat H_r(y)\right)\right)\Big/\left(E\delta\left(\int_0^\infty \bar F_1(y)\bar F_2(y)\,d\hat H_r(y)\right.\right.$$
$$\left.\left.+\int_0^\infty \bar\Phi_1(y)\bar F_2(y)\,d\hat H_r(y)+\int_0^\infty \bar\Phi_2(y)\bar F_1(y)\,d\hat H_r(y)\right)\right)=$$
$$=\frac{(c_1+c_2)\left(E(\alpha_1\wedge\alpha_2)+\int_0^\infty \bar F_2(t)\bar\Phi_1(t)\,dt+\int_0^\infty \bar F_1(t)\bar\Phi_2(t)\,dt\right)}{E\delta\left(\int_0^\infty \bar F_1(y)\bar F_2(y)\,d\hat H_r(y)+\int_0^\infty \bar\Phi_1(y)\bar F_2(y)\,d\hat H_r(y)+\int_0^\infty \bar\Phi_2(y)\bar F_1(y)\,d\hat H_r(y)\right)}-c_2,$$

and for average specific expenses the following formula is true:

$$C=\left(c_2E\delta\left(\int_0^\infty \bar F_1(y)\bar F_2(y)\,d\hat H_r(y)+\int_0^\infty \bar\Phi_1(y)\bar F_2(y)\,d\hat H_r(y)+\int_0^\infty \bar\Phi_2(y)\bar F_1(y)\,d\hat H_r(y)\right)-\right.$$
$$\left.-c_2\left(E(\alpha_1\wedge\alpha_2)+\int_0^\infty \bar F_2(t)\bar\Phi_1(t)\,dt+\int_0^\infty \bar F_1(t)\bar\Phi_2(t)\,dt\right)\right)\Big/\left(E(\alpha_1\wedge\alpha_2)+\right.$$
$$\left.+\int_0^\infty \bar F_2(t)\bar\Phi_1(t)\,dt+\int_0^\infty \bar F_1(t)\bar\Phi_2(t)\,dt\right)=$$
$$=\frac{c_2E\delta\left(\int_0^\infty \bar F_1(y)\bar F_2(y)\,d\hat H_r(y)+\int_0^\infty \bar\Phi_1(y)\bar F_2(y)\,d\hat H_r(y)+\int_0^\infty \bar\Phi_2(y)\bar F_1(y)\,d\hat H_r(y)\right)}{E(\alpha_1\wedge\alpha_2)+\int_0^\infty \bar F_2(t)\bar\Phi_1(t)\,dt+\int_0^\infty \bar F_1(t)\bar\Phi_2(t)\,dt}-c_2.$$

Let us consider the homogeneous case, DF of components κ_1, κ_2 times to failure being the same: $F_1(t)=F_2(t)=F(t), f_1(t)=f_2(t)=f(t)$.

Then, due to the states symmetry, we get:

$$\varphi_2(x,z)=\varphi_3(x,z),\varphi_4(x)=\varphi_5(x),\varphi_6(z)=\varphi_7(z).$$

The system (3.7) can be reduced to the following one:

$$
\begin{cases}
\rho_0 = 2\int_0^\infty \varphi_6(z)\,dz, \\
\varphi_1(x_1,x_2) = \rho_0 \int_0^\infty f(x_1+t)f(x_2+t)r(t)dt + \int_0^\infty \varphi_1(x_1+t,x_2+t)r(t)dt + \\
+\int_0^\infty \varphi_4(x_2+t)f(x_1+t)r(t)dt + \int_0^\infty \varphi_4(x_1+t)f(x_2+t)r(t)dt, \\
\varphi_2(x,z) = \rho_0 \int_0^\infty f(t)f(x+t)r(z+t)dt + \int_0^\infty \varphi_4(x+t)f(t)r(z+t)dt + \\
+\int_0^\infty \varphi_4(t)f(x+t)r(z+t)dt + \int_0^\infty \varphi_1(t,x+t)r(z+t)dt, \\
\varphi_4(x) = \int_0^\infty \varphi_2(x+t,t)dt, \\
\varphi_6(z) = \int_0^\infty \varphi_2(t,z+t)dt, \\
\rho_0 + \int_0^\infty\int_0^\infty \varphi_1(x_1,x_2)dx_1\,dx_2 + 2\int_0^\infty\int_0^\infty \varphi_{21}(x,z)\,dx\,dz + 2\int_0^\infty \varphi_4(x)dx + 2\int_0^\infty \varphi_6(z)dz = 1.
\end{cases}
\tag{3.20}
$$

The solution of the system (3.20) is as follows:

$$\varphi_4(x) = \varphi_5(x) = \rho_0 \psi(x),$$

where

$$\psi(x) = \int_0^\infty f(x+y)h(y)\,dy + \int_0^\infty \pi(x,y)\,dy \int_0^\infty f(y+t)h(t)\,dt,$$

$$\varphi_1(x_1,x_2) = \rho_0 \left(\int_0^\infty f(x_1+y)f(x_2+y)h_r(y)dy + \right.$$

$$\left. + \int_0^\infty \psi(x_1+t)f(x_2+t)h_r(t)\,dt + \int_0^\infty \psi(x_2+t)f(x_1+t)h_r(t)\,dt \right),$$

$$\varphi_2(x,z) = \varphi_3(x,z) = \rho_0 \left(\int_0^\infty f(y)f(x+y)v_r(y,z)dy + \right.$$

$$\left. + \int_0^\infty \psi(t)f(x+t)v_r(t,z)\,dt + \int_0^\infty \psi(x+t)f(t)v_r(t,z)\,dt \right),$$

$$\varphi_6(z) = \varphi_7(z) = \rho_0 \left(\int_0^\infty dt \int_0^\infty f(y)f(t+y)v_r(y,t+z)dy + \right.$$

$$\left. + \int_0^\infty dt \int_0^\infty \psi(y)f(t+y)v_r(y,t+z)\,dy + \int_0^\infty dt \int_0^\infty \psi(t+y)f(y)v_r(y,t+z)\,dy \right),$$

where $h_r(t)=\sum_{n=1}^{\infty} r^{*(n)}(t)$ is the density of renewal function $H_r(t)$ of the renewal process generated by RV δ; $v_r(z,x)=r(z+x)+\int_0^z r(z+x-s)h_r(s)\,ds$ is the distribution density of the direct residual time for the renewal process generated by RV δ; $h(t)=\sum_{n=1}^{\infty}\tilde{\gamma}^{*(n)}(t)$ is the density of renewal function of the renewal process generates by RV with densities $\tilde{\gamma}(t)=\int_0^t f(t)v_r(y,t-y)\,dy$;

$$\gamma(x,y)=\int_0^\infty f(x+y+z)v_r(y,z)\,dz+\int_0^\infty h(s)\,ds\int_0^\infty f(x+y+s+z)v_r(y,z)\,dz;$$

$$\pi(x,y)=\sum_{n=1}^{\infty}\gamma^{(n)}(x,y),\gamma^{(1)}(x,y)=\gamma(x,y),\gamma^{(n)}(x,y)=\int_0^\infty \gamma(x,t)\gamma^{(n-1)}(t,y)\,dt.$$

Let us define system stationary characteristics:

$$\int_{E+} m(e)\rho(de)=\rho_0\left(E(\alpha_1\wedge\alpha_2)+2\int_0^\infty \bar{F}(t)\bar{\Phi}(t)\,dt\right),$$

where

$$\bar{\Phi}(t)=\int_0^\infty \bar{F}(t+y)h(y)\,dy+\int_0^\infty \bar{\Gamma}(t,y)\,dy\int_0^\infty f(y+z)h(z)\,dz+$$

$$+\int_0^\infty \bar{\Pi}(t,y)\,dy\int_0^\infty f(y+z)h(z)\,dz+\int_0^\infty \bar{\Pi}(t,y)\,dy\int_0^\infty \gamma(y,z)\,dz\int_0^\infty f(z+s)h(s)\,ds,$$

$$\bar{\Gamma}(t,y)=\int_t^\infty \gamma(x,y)\,dx,\bar{\Pi}(t,y)=\int_t^\infty \pi(x,y)\,dx;$$

$$\int_{E_-} m(e)\rho(de)=\rho_0\left(E\delta\int_0^\infty\left(\bar{F}(y)\right)^2 d\hat{H}_r(y)+2E\delta\int_0^\infty \bar{\Phi}(y)\bar{F}(y)\,d\hat{H}_r(y)-\right.$$

$$\left.-E\left(\alpha_1\wedge\alpha_2\right)-2\int_0^\infty \bar{\Phi}(y)\bar{F}(y)\,dy\right);$$

$$\int_{E_+} P(e,E_-)\rho(de)=\rho_0\left(1+2\bar{\Phi}(0)\right),$$

where

$$\bar{\Phi}(0)=\int_0^\infty \bar{F}(y)h(y)\,dy+\int_0^\infty \bar{\Gamma}(0,y)\,dy\int_0^\infty f(y+z)h(z)\,dz+$$

$$+\int_0^\infty \bar{\Pi}(0,y)\,dy\int_0^\infty f(y+z)h(z)\,dz+\int_0^\infty \bar{\Pi}(0,y)\,dy\int_0^\infty \gamma(y,z)\,dz\int_0^\infty f(z+s)h(s)\,ds,$$

$$\bar{\Gamma}(0,y)=\int_0^\infty \gamma(x,y)\,dx,\bar{\Pi}(0,y)=\int_0^\infty \pi(x,y)\,dx.$$

In such a way, average stationary operating TF T_+ and average stationary restoration time T_- are respectively:

$$T_+ = \frac{E(\alpha_1 \wedge \alpha_2) + 2\int_0^\infty \bar{\Phi}(t)\bar{F}(t)\,dt}{1 + 2\bar{\Phi}(0)} \text{ and}$$

$$T_- = \frac{E\delta\int_0^\infty \left(\bar{F}(y)\right)^2 d\hat{H}_r(y) + 2E\delta\int_0^\infty \bar{\Phi}(y)\bar{F}(y)\,d\hat{H}_r(y) - E\left(\alpha_1 \wedge \alpha_2\right) - 2\int_0^\infty \bar{\Phi}(y)\bar{F}(y)\,dy}{1 + 2\bar{\Phi}(0)}.$$

Stationary availability factor is given by the following formula:

$$K_a = \frac{E(\alpha_1 \wedge \alpha_2) + 2\int_0^\infty \bar{F}(t)\bar{\Phi}(t)\,dt}{E\delta\left(\int_0^\infty \left(\bar{F}(y)\right)^2 d\hat{H}_r(y) + 2\int_0^\infty \bar{\Phi}(y)\bar{F}(y)\,d\hat{H}_r(y)\right)}.$$

Consider the case of RV α_i, $i = 1, 2$, δ exponential distribution: $F_i(t) = 1 - e^{-\lambda_i t}$, $i = 1,2$, $R(t) = 1 - e^{-\mu t}$.

Then the solution of the system of equations (3.5) is:

$$\rho_0 = \rho(3111), \rho(3111x_1x_2) = \rho_0 \frac{\mu(\lambda_1 + \mu)(\lambda_2 + \mu)}{\lambda_1 + \lambda_2 + 2\mu} e^{-\lambda_1 x_1} e^{-\lambda_2 x_2},$$

$$\rho(1011xz) = \rho_0 \frac{\mu(\lambda_1 + \mu)(\lambda_2 + \mu)}{\lambda_1 + \lambda_2 + 2\mu} e^{-\lambda_2 x} e^{-\mu z},$$

$$\rho(2101xz) = \rho_0 \frac{\mu(\lambda_1 + \mu)(\lambda_2 + \mu)}{\lambda_1 + \lambda_2 + 2\mu} e^{-\lambda_1 x} e^{-\mu z},$$

$$\rho(1111x) = \rho_0 \frac{\mu(\lambda_1 + \mu)}{\lambda_1 + \lambda_2 + 2\mu} e^{-\lambda_2 x},\ \rho(2111x) = \rho_0 \frac{\mu(\lambda_2 + \mu)}{\lambda_1 + \lambda_2 + 2\mu} e^{-\lambda_1 x},$$

$$\rho(1001z) = \rho_0 \frac{\mu(\lambda_2 + \mu)}{\lambda_1 + \lambda_2 + 2\mu} e^{-\mu z},\ \rho(2001z) = \rho_0 \frac{\mu(\lambda_1 + \mu)}{\lambda_1 + \lambda_2 + 2\mu} e^{-\mu z}.$$

Let us obtain system stationary characteristics for this case.

$$\int_{E+} m(e)\rho(de) = \rho_0 \frac{(\lambda_1 + \mu)(\lambda_2 + \mu)}{\lambda_1\lambda_2(\lambda_1 + \lambda_2 + 2\mu)},\ \int_{E_-} m(e)\rho(de) = \rho_0 \frac{(\lambda_1 + \mu)(\lambda_2 + \mu)(\lambda_1 + \lambda_2)}{\lambda_1\lambda_2\mu(\lambda_1 + \lambda_2 + 2\mu)},$$

$$\int_{E_+} P(e, E_-)\rho(de) = \rho_0 \frac{(\lambda_1 + \mu)(\lambda_2 + \mu)(\lambda_1 + \lambda_2)}{\lambda_1\lambda_2(\lambda_1 + \lambda_2 + 2\mu)}.$$

In such a way, average stationary operating TF T_+ and average stationary restoration time T_- are:

$$T_+ = \frac{E\alpha_1 E\alpha_2}{E\alpha_1 + E\alpha_2}, T_- = E\delta.$$

Stationary availability factor, average specific income per calendar time unit, and average specific expenses per time unit of up-state look like:

$$\kappa_a = \frac{E\alpha_1 E\alpha_2}{E\alpha_1 E\alpha_2 + E\alpha_1 E\delta + E\alpha_2 E\delta},$$

$$S = \frac{c_1 E\alpha_1 E\alpha_2 - c_2 E\delta(E\alpha_1 + E\alpha_2)}{E\alpha_1 E\alpha_2 + E\alpha_1 E\delta + E\alpha_2 E\delta}, C = \frac{c_2 E\delta(E\alpha_1 + E\alpha_2)}{E\alpha_1 E\alpha_2}.$$

Now consider the case of RV α_1, α_2 exponential distributions and nonrandom control periodicity $\tau > 0$. Taking into account that $F_i(t) = 1 - e^{-\lambda_i t}$ where $i = 1,2$, and $R(t) = 1(t - \tau)$, $E\delta = \tau$, the solution of the system (3.5) takes the following form:

$$\begin{cases} \rho_0 = \rho(3111), \quad \rho(3111x_1x_2) = \rho_0 \dfrac{\lambda_1\lambda_2}{(e^{\lambda_1\tau} - 1)(e^{\lambda_2\tau} - 1)} e^{-\lambda_1 x_1} e^{-\lambda_2 x_2}, \\ \rho(1011xz) = \rho_0 \dfrac{\lambda_1\lambda_2}{(e^{\lambda_1\tau} - 1)(e^{\lambda_2\tau} - 1)} e^{-\lambda_2 x} e^{(\lambda_1+\lambda_2)z}, \\ \rho(2101xz) = \rho_0 \dfrac{\lambda_1\lambda_2}{(e^{\lambda_1\tau} - 1)(e^{\lambda_2\tau} - 1)} e^{-\lambda_1 x} e^{(\lambda_1+\lambda_2)z}, \\ \rho(1111x) = \rho_0 \dfrac{\lambda_2}{(e^{\lambda_2\tau} - 1)} e^{-\lambda_2 x}, \quad \rho(2111x) = \rho_0 \dfrac{\lambda_1}{(e^{\lambda_1\tau} - 1)} e^{-\lambda_1 x}, \\ \rho(1001z) = \rho_0 \dfrac{\lambda_1}{(e^{\lambda_1\tau} - 1)(e^{\lambda_2\tau} - 1)} e^{\lambda_1 z}(e^{\lambda_2\tau} - e^{\lambda_2 z}), \\ \rho(2001z) = \rho_0 \dfrac{\lambda_2}{(e^{\lambda_1\tau} - 1)(e^{\lambda_2\tau} - 1)} e^{\lambda_2 z}(e^{\lambda_1\tau} - e^{\lambda_1 z}). \end{cases}$$

Average stationary operating TF T_+ and average stationary restoration time T_- are:

$$T_+ = \frac{1}{\lambda_1 + \lambda_2} = \frac{E\alpha_1 E\alpha_2}{E\alpha_1 + E\alpha_2}, T_- = \frac{(\lambda_1 + \lambda_2)\tau e^{(\lambda_1+\lambda_2)\tau} - e^{(\lambda_1+\lambda_2)\tau} + 1}{(\lambda_1 + \lambda_2)\left(e^{(\lambda_1+\lambda_2)\tau} - 1\right)}.$$

We get the expression for the stationary availability factor:

$$K_a = \frac{e^{(\lambda_1+\lambda_2)\tau} - 1}{(\lambda_1 + \lambda_2)\tau e^{(\lambda_1+\lambda_2)\tau}} = \frac{E\alpha_1 E\alpha_2\left(1 - e^{-(\lambda_1+\lambda_2)\tau}\right)}{(E\alpha_1 + E\alpha_2)\tau}. \tag{3.21}$$

TABLE 3.1 Values of $K_a(\tau)$, $S(\tau)$, $C(\tau)$ Under $\tau = 5$ h

Initial data		Calculation results		
$E\alpha_1$, h	$E\alpha_2$, h	$K_a(\tau)$	$S(\tau)$, c.u./h	$C(\tau)$, c.u./h
90	70	0.9395	2.698	0.129
90	50	0.9266	2.633	0.158
90	10	0.7687	1.844	0.601

Average specific income per calendar time unit and average specific expenses per time unit of up-state can be defined by the ratios:

$$S = \frac{(c_1 + c_2)\left(1 - e^{-(\lambda_1+\lambda_2)\tau}\right)}{(\lambda_1 + \lambda_2)\tau} - c_2 = \frac{E\alpha_1 E\alpha_2 (c_1 + c_2)\left(1 - e^{-(\lambda_1+\lambda_2)\tau}\right)}{\tau(E\alpha_1 + E\alpha_2)} - c_2,$$
$$C = \frac{c_2(\lambda_1 + \lambda_2)\tau}{\left(1 - e^{-(\lambda_1+\lambda_2)\tau}\right)} - c_2 = \frac{c_2\tau(E\alpha_1 + E\alpha_2)}{\left(1 - e^{-(\lambda_1+\lambda_2)\tau}\right)E\alpha_1 E\alpha_2} - c_2. \tag{3.22}$$

Example. Initial data and calculation results are given in Table 3.1 under the values $c_1 = 3$ c.u., $c_2 = 2$ c.u.

The graphs of functions $K_a(\tau)$, $S(\tau)$, $C(\tau)$ for the exponential distributions are given in Figures 3.3, 3.4, and 3.5, respectively.

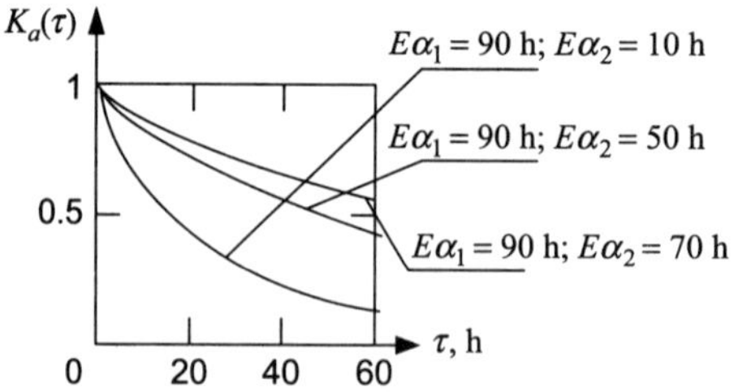

FIGURE 3.3 Graph of stationary availability factor $K_a(\tau)$ against control periodicity τ

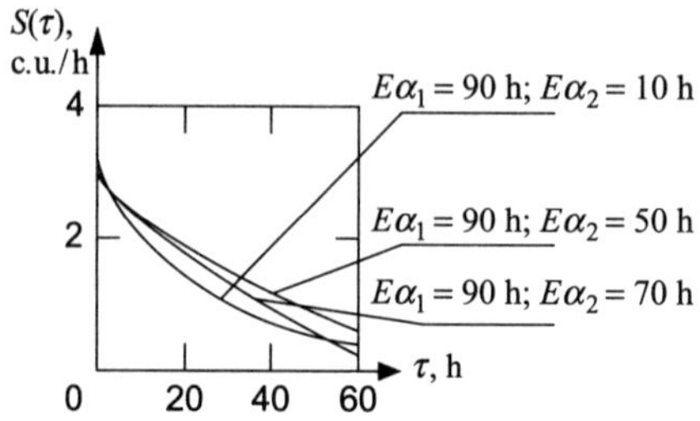

FIGURE 3.4 Graph of average specific income $S(\tau)$ against control periodicity τ

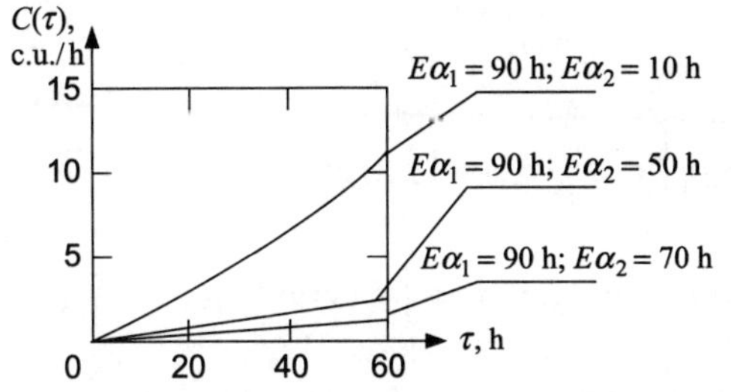

FIGURE 3.5 Graph of average specific expenses $C(\tau)$ against control periodicity τ

3.2 THE MODEL OF TWO-COMPONENT PARALLEL SYSTEM WITH IMMEDIATE CONTROL AND RESTORATION

3.2.1 System Description

In this section, stationary characteristics of the system described in Section 3.1, in case of components κ_1 and κ_2 parallel connection (in reliability sense) are defined. System transition graph, with regard to parallel connection of κ_1 and κ_2, is given in Figure 3.6.

3.2.2 Definition of System Stationary Characteristics

In this case, system phase state space E is split into two subsets:

$$E_+ = \{3111,\ \ 3111x_1x_2,\ \ 1111x,\ \ 2111x,\ \ 1011xz,\ \ 2101xz\} - \text{system up-state};$$

$E_- = \{2001z, 1001z\}$ – system down-state.

Let us find average stationary operating TF T_+ and average stationary restoration time T_- with the help of formulas given in Chapter 1 (1.17 and 1.18).

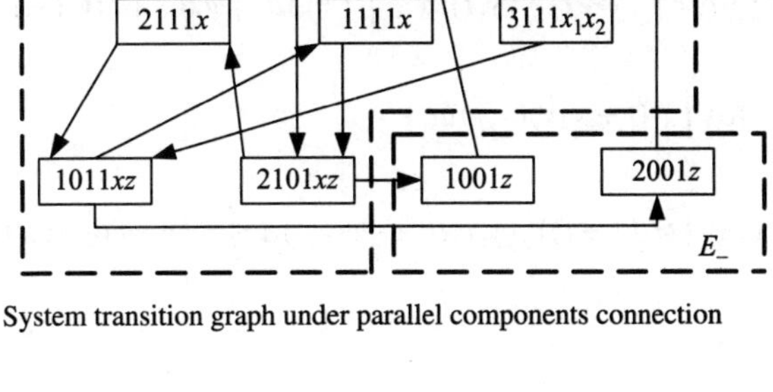

FIGURE 3.6 System transition graph under parallel components connection

Then,

$$\int_{E+} m(e)\rho(de) = \rho_0 \Bigg(E(\alpha_1 \wedge \alpha_2) + \int_0^\infty \bar{F}_2(t)\bar{\Phi}_1(t)\,dt + \int_0^\infty \bar{F}_1(t)\bar{\Phi}_2(t)\,dt +$$
$$+\int_0^\infty f_1(y)\,dy \int_0^\infty \bar{F}_2(x+y)\bar{V}_r(y,x)\,dx + \int_0^\infty f_2(y)\,dy \int_0^\infty \bar{F}_1(x+y)\bar{V}_r(y,x)\,dx + \tag{3.23}$$
$$+\int_0^\infty \psi_1(y)\,dy \int_0^\infty \bar{F}_2(x+y)\bar{V}_r(y,x)\,dx + \int_0^\infty f_2(y)\,dy \int_0^\infty \bar{\Phi}_1(x+y)\bar{V}_r(y,x)\,dx +$$
$$+\int_0^\infty f_1(y)\,dy \int_0^\infty \bar{\Phi}_2(x+y)\bar{V}_r(y,x)\,dx + \int_0^\infty \psi_2(y)\,dy \int_0^\infty \bar{F}_1(x+y)\bar{V}_r(y,x)\,dx \Bigg),$$

where $V_r(y,x)$ is the DF of the direct residual time for the renewal process, generated by RV δ, $\bar{V}_r(y,x) = 1 - V_r(y,x)$.

$$\int_{E\_} m(e)\rho(de) = \rho_0 \Bigg(E\delta \int_0^\infty \bar{F}_1(y)\bar{F}_2(y)\,d\hat{H}_r(y) + E\delta \int_0^\infty \bar{\Phi}_1(y)\bar{F}_2(y)\,d\hat{H}_r(y) +$$
$$+E\delta \int_0^\infty \bar{\Phi}_2(y)\bar{F}_1(y)\,d\hat{H}_r(y) - E(\alpha_1 \wedge \alpha_2) - \int_0^\infty \bar{\Phi}_1(y)\bar{F}_2(y)\,dy - \int_0^\infty \bar{\Phi}_2(y)\bar{F}_1(y)\,dy -$$
$$-\int_0^\infty f_1(y)\,dy \int_0^\infty \bar{V}_r(y,z)\bar{F}_2(z+y)\,dz - \int_0^\infty f_2(y)\,dy \int_0^\infty \bar{V}_r(y,z)\bar{F}_1(z+y)\,dz - \tag{3.24}$$
$$-\int_0^\infty \psi_1(y)\,dy \int_0^\infty \bar{V}_r(y,z)\bar{F}_2(z+y)\,dz - \int_0^\infty f_2(y)\,dy \int_0^\infty \bar{V}_r(y,z)\bar{\Phi}_1(z+y)\,dz -$$
$$-\int_0^\infty \psi_2(y)\,dy \int_0^\infty \bar{V}_r(y,z)\bar{F}_1(z+y)\,dz - \int_0^\infty f_1(y)\,dy \int_0^\infty \bar{V}_r(y,z)\bar{\Phi}_2(z+y)\,dz \Bigg).$$

Next,

$$\int_{E_+} P(e,E\_)\rho(de) = \rho_0 \Bigg(\int_0^\infty f_1(y)dy \int_0^\infty f_2(t+y)\bar{V}_r(y,t)\,dt +$$
$$+\int_0^\infty \psi_1(y)\,dy \int_0^\infty f_2(y+t)\bar{V}_r(y,t)\,dt + \int_0^\infty f_1(y)\,dy \int_0^\infty \psi_2(y+t)\bar{V}_r(y,t)\,dt +$$
$$+\int_0^\infty f_2(y)\,dy \int_0^\infty f_1(t+y)\bar{V}_r(y,t)\,dt +$$
$$+\int_0^\infty f_2(y)\,dy \int_0^\infty \psi_1(y+t)\bar{V}_r(y,t)\,dt + \int_0^\infty \psi_2(t)\,dy \int_0^\infty f_1(y+t)\bar{V}_r(y,t)\,dt \Bigg).$$

One can prove that:

$$\int_0^\infty f_1(y)dy\int_0^\infty f_2(t+y)\bar{V}_r(y,t)\,dt+\int_0^\infty \psi_1(y)\,dy\int_0^\infty f_2(y+t)\bar{V}_r(y,t)\,dt+$$
$$+\int_0^\infty f_1(y)\,dy\int_0^\infty \psi_2(y+t)\bar{V}_r(y,t)\,dt+\int_0^\infty f_2(y)\,dy\int_0^\infty f_1(t+y)\bar{V}_r(y,t)\,dt+$$
$$+\int_0^\infty f_2(y)\,dy\int_0^\infty \psi_1(y+t)\bar{V}_r(y,t)\,dt+\int_0^\infty \psi_2(t)\,dy\int_0^\infty f_1(y+t)\bar{V}_r(y,t)\,dt=1.$$

Consequently,

$$\int_{E_+} P(e,E_-)\rho(de)=\rho_0. \tag{3.25}$$

In such a way, taking into account (3.23) and (3.25), we get the expression for the average stationary operation TF T_+:

$$\begin{aligned}T_+ &= E(\alpha_1\wedge\alpha_2)+\int_0^\infty \bar{F}_2(t)\bar{\Phi}_1(t)\,dt+\int_0^\infty \bar{F}_1(t)\bar{\Phi}_2(t)\,dt+\\
&+\int_0^\infty f_1(y)\,dy\int_0^\infty \bar{F}_2(x+y)\bar{V}_r(y,x)\,dx+\int_0^\infty f_2(y)\,dy\int_0^\infty \bar{F}_1(x+y)\bar{V}_r(y,x)\,dx+\\
&+\int_0^\infty \psi_1(y)\,dy\int_0^\infty \bar{F}_2(x+y)\bar{V}_r(y,x)\,dx+\int_0^\infty f_2(y)\,dy\int_0^\infty \bar{\Phi}_1(x+y)\bar{V}_r(y,x)\,dx+\\
&+\int_0^\infty f_1(y)\,dy\int_0^\infty \bar{\Phi}_2(x+y)\bar{V}_r(y,x)\,dx+\int_0^\infty \psi_2(y)\,dy\int_0^\infty \bar{F}_1(x+y)\bar{V}_r(y,x)\,dx.\end{aligned} \tag{3.26}$$

Average stationary restoration time T_-, according to (3.24) and (3.25), is:

$$\begin{aligned}T_- &= E\delta\int_0^\infty \bar{F}_1(y)\bar{F}_2(y)\,d\hat{H}_r(y)+E\delta\int_0^\infty \bar{\Phi}_1(y)\bar{F}_2(y)\,d\hat{H}_r(y)+\\
&+E\delta\int_0^\infty \bar{\Phi}_2(y)\bar{F}_1(y)\,d\hat{H}_r(y)-E(\alpha_1\wedge\alpha_2)-\int_0^\infty \bar{\Phi}_1(y)\bar{F}_2(y)\,dy-\int_0^\infty \bar{\Phi}_2(y)\bar{F}_1(y)\,dy-\\
&-\int_0^\infty f_1(y)\,dy\int_0^\infty \bar{V}_r(y,z)\bar{F}_2(z+y)\,dz-\int_0^\infty f_2(y)\,dy\int_0^\infty \bar{V}_r(y,z)\bar{F}_1(z+y)\,dz-\\
&-\int_0^\infty \psi_1(y)\,dy\int_0^\infty \bar{V}_r(y,z)\bar{F}_2(z+y)\,dz-\int_0^\infty f_2(y)\,dy\int_0^\infty \bar{V}_r(y,z)\bar{\Phi}_1(z+y)\,dz-\\
&-\int_0^\infty \psi_2(y)\,dy\int_0^\infty \bar{V}_r(y,z)\bar{F}_1(z+y)\,dz.\end{aligned} \tag{3.27}$$

Stationary availability factor, with regard to formulas (3.26) and (3.27), is defined as follows:

$$K_a = \left(E(\alpha_1 \wedge \alpha_2) + \int_0^\infty \overline{F}_2(t)\overline{\Phi}_1(t)\,dt + \int_0^\infty \overline{F}_1(t)\overline{\Phi}_2(t)\,dt + \right.$$
$$+ \int_0^\infty f_1(y)\,dy \int_0^\infty \overline{F}_2(x+y)\overline{V}_r(y,x)\,dx + \int_0^\infty f_2(y)\,dy \int_0^\infty \overline{F}_1(x+y)\overline{V}_r(y,x)\,dx +$$
$$+ \int_0^\infty \psi_1(y)\,dy \int_0^\infty \overline{F}_2(x+y)\overline{V}_r(y,x)\,dx + \int_0^\infty f_2(y)\,dy \int_0^\infty \overline{\Phi}_1(x+y)\overline{V}_r(y,x)\,dx +$$
$$+ \int_0^\infty f_1(y)\,dy \int_0^\infty \overline{\Phi}_2(x+y)\overline{V}_r(y,x)\,dx +$$
$$\left. \int_0^\infty \psi_2(y)\,dy \int_0^\infty \overline{F}_1(x+y)\overline{V}_r(y,x)\,dx \right) \Bigg/ \left(E\delta \left(\int_0^\infty \overline{F}_1(y)\overline{F}_2(y)\,d\hat{H}_r(y) + \right. \right.$$
$$\left. \left. \int_0^\infty \overline{\Phi}_1(y)\overline{F}_2(y)\,d\hat{H}_r(y) + \int_0^\infty \overline{\Phi}_2(y)\overline{F}_1(y)\,d\hat{H}_r(y) \right) \right).$$

We can find average specific income S per calendar time unit and average specific expenses C per time unit of up-state by means of formulas (1.22) and (1.23).

Let c_1 be the income time unit of up-state; c_2 be expenses for the system failure per time unit.

For the system studied $f_s(e), f_c(e)$ are:

$$f_s(e) = \begin{cases} c_1, & e \in \{3111,\ 3111x_1x_2,\ 1011x\ z,\ 2101x\ z,\ 1111x,\ 2111x\}, \\ -c_2, & e \in \{1001z,\ 2001z\}, \end{cases}$$
$$f_c(e) = \begin{cases} 0, & e \in \{3111,\ 3111x_1x_2,\ 1011x\ z,\ 2101x\ z,\ 1111x,\ 2111x\}, \\ c_2, & e \in \{1001z,\ 2001z\}. \end{cases} \tag{3.28}$$

With regard to (3.28), average income is defined by the expression:

$$S = \left((c_1 + c_2) \left(E(\alpha_1 \wedge \alpha_2) + \int_0^\infty \overline{F}_2(t)\overline{\Phi}_1(t)\,dt + \int_0^\infty \overline{F}_1(t)\overline{\Phi}_2(t)\,dt + \right. \right.$$
$$+ \int_0^\infty f_1(y)\,dy \int_0^\infty \overline{F}_2(x+y)\overline{V}_r(y,x)\,dx + \int_0^\infty f_2(y)\,dy \int_0^\infty \overline{F}_1(x+y)\overline{V}_r(y,x)\,dx +$$
$$\int_0^\infty \psi_1(y)\,dy \int_0^\infty \overline{F}_2(x+y)\overline{V}_r(y,x)\,dx + \int_0^\infty f_2(y)\,dy \int_0^\infty \overline{\Phi}_1(x+y)\overline{V}_r(y,x)\,dx + \tag{3.29}$$
$$\left. + \int_0^\infty f_1(y)\,dy \int_0^\infty \overline{\Phi}_2(x+y)\overline{V}_r(y,x)\,dx + \int_0^\infty \psi_2(y)\,dy \int_0^\infty \overline{F}_1(x+y)\overline{V}_r(y,x)\,dx \right) -$$

$$-c_2E\delta\left(\int_0^\infty \bar{F}_1(y)\bar{F}_2(y)\,d\hat{H}_r(y)+\int_0^\infty \bar{\Phi}_1(y)\bar{F}_2(y)\,d\hat{H}_r(y)+\right.$$

$$\left.\left.\int_0^\infty \bar{\Phi}_2(y)\bar{F}_1(y)\,d\hat{H}_r(y)\right)\right)\Bigg/\left(E\delta\left(\int_0^\infty \bar{F}_1(y)\bar{F}_2(y)\,d\hat{H}_r(y)+\right.\right.$$

$$\left.\left.\int_0^\infty \bar{\Phi}_1(y)\bar{F}_2(y)\,d\hat{H}_r(y)+\int_0^\infty \bar{\Phi}_2(y)\bar{F}_1(y)\,d\hat{H}_r(y)\right)\right).$$

Taking into account (3.26), we get the expression (3.29):

$$S=\frac{(c_1+c_2)T_+}{E\delta\left(\int_0^\infty \bar{F}_1(y)\bar{F}_2(y)\,d\hat{H}_r(y)+\int_0^\infty \bar{\Phi}_1(y)\bar{F}_2(y)\,d\hat{H}_r(y)+\int_0^\infty \bar{\Phi}_2(y)\bar{F}_1(y)\,d\hat{H}_r(y)\right)}-c_2.$$

Average expenses are given by:

$$\begin{aligned}C=&\left(c_2E\delta\left(\int_0^\infty \bar{F}_1(y)\bar{F}_2(y)\,d\hat{H}_r(y)+\int_0^\infty \bar{\Phi}_1(y)\bar{F}_2(y)\,d\hat{H}_r(y)+\int_0^\infty \bar{\Phi}_2(y)\bar{F}_1(y)\,d\hat{H}_r(y)\right)-\right.\\
&-c_2\left(E(\alpha_1\wedge\alpha_2)+\int_0^\infty \bar{F}_2(t)\bar{\Phi}_1(t)\,dt+\int_0^\infty \bar{F}_1(t)\bar{\Phi}_2(t)\,dt+\int_0^\infty f_1(y)\,dy\int_0^\infty \bar{F}_2(x+y)\bar{V}_r(y,x)\,dx+\right.\\
&+\int_0^\infty f_2(y)\,dy\int_0^\infty \bar{F}_1(x+y)\bar{V}_r(y,x)\,dx+\int_0^\infty \psi_1(y)\,dy\int_0^\infty \bar{F}_2(x+y)\bar{V}_r(y,x)\,dx+\\
&+\int_0^\infty f_2(y)\,dy\int_0^\infty \bar{\Phi}_1(x+y)\bar{V}_r(y,x)\,dx+\int_0^\infty f_1(y)\,dy\int_0^\infty \bar{\Phi}_2(x+y)\bar{V}_r(y,x)\,dx+\\
&\left.\left.+\int_0^\infty \psi_2(y)\,dy\int_0^\infty \bar{F}_1(x+y)\bar{V}_r(y,x)\,dx\right)\right)\Bigg/\left(E(\alpha_1\wedge\alpha_2)+\int_0^\infty \bar{F}_2(t)\bar{\Phi}_1(t)\,dt+\right.\\
&+\int_0^\infty \bar{F}_1(t)\bar{\Phi}_2(t)\,dt+\int_0^\infty f_1(y)\,dy\int_0^\infty \bar{F}_2(x+y)\bar{V}_r(y,x)\,dx+\int_0^\infty f_2(y)\,dy\int_0^\infty \bar{F}_1(x+y)\bar{V}_r(y,x)\,dx+\\
&+\int_0^\infty \psi_1(y)\,dy\int_0^\infty \bar{F}_2(x+y)\bar{V}_r(y,x)\,dx+\int_0^\infty f_2(y)\,dy\int_0^\infty \bar{\Phi}_1(x+y)\bar{V}_r(y,x)\,dx+\\
&\left.+\int_0^\infty f_1(y)\,dy\int_0^\infty \bar{\Phi}_2(x+y)\bar{V}_r(y,x)\,dx+\int_0^\infty \psi_2(y)\,dy\int_0^\infty \bar{F}_1(x+y)\bar{V}_r(y,x)\,dx\right).\end{aligned} \tag{3.30}$$

With regard to (3.26), the formula (3.30) takes the form:

$$C=\frac{c_2E\delta\left(\int_0^\infty \bar{F}_1(y)\bar{F}_2(y)\,d\hat{H}_r(y)+\int_0^\infty \bar{\Phi}_1(y)\bar{F}_2(y)\,d\hat{H}_r(y)+\int_0^\infty \bar{\Phi}_2(y)\bar{F}_1(y)\,d\hat{H}_r(y)\right)}{T_+}-c_2.$$

Consider the case of RV α_1, α_2, δ exponential distributions: $F_i(t)=1-e^{-\lambda_i t}$, where i = 1,2, $R(t)=1-e^{-\mu t}$.

Let us write out system stationary characteristics.

$$\int_{E+} m(e)\rho(de) = \rho_0 \frac{(\lambda_1+\lambda_2+\mu)^2-\lambda_1\lambda_2}{\lambda_1\lambda_2(\lambda_1+\lambda_2+2\mu)}, \int_{E_-} m(e)\rho(de) = \rho_0 \frac{1}{\mu} = \rho_0 E\delta,$$

$$\int_{E_+} P(e,E_-)\rho(de) = \rho_0.$$

Thus, average stationary operation TF T_+ and average stationary restoration time T_- are:

$$T_+ = \frac{(\lambda_1+\lambda_2+\mu)^2-\lambda_1\lambda_2}{\lambda_1\lambda_2(\lambda_1+\lambda_2+2\mu)}, T_- = E\delta.$$

Stationary availability factor is:

$$K_a = \frac{\mu\left((\lambda_1+\lambda_2+\mu)^2-\lambda_1\lambda_2\right)}{(\lambda_1+\mu)(\lambda_2+\mu)(\lambda_1+\lambda_2+\mu)}.$$

Average specific income and average specific expenses can be given by:

$$S = \frac{\mu(c_1+c_2)\left((\lambda_1+\lambda_2+\mu)^2-\lambda_1\lambda_2\right)}{(\lambda_1+\mu)(\lambda_2+\mu)(\lambda_1+\lambda_2+\mu)} - c_2, C = \frac{c_2\lambda_1\lambda_2(\lambda_1+\lambda_2+2\mu)}{\mu\left((\lambda_1+\lambda_2+\mu)^2-\lambda_1\lambda_2\right)} - c_2.$$

Consider the case of nonrandom control periodicity $\tau > 0$: $R(t) = 1(t-\tau)$, $E\delta = \tau$.

In this case, average stationary operating TF T_+ and average stationary restoration time T_- are determined by the following formulas:

$$T_+ = \frac{\lambda_1^2 e^{\lambda_1\tau}\left(e^{\lambda_2\tau}-1\right)+\lambda_2^2 e^{\lambda_2\tau}\left(e^{\lambda_1\tau}-1\right)+\lambda_1\lambda_2\left(e^{\lambda_1\tau}-1\right)\left(e^{\lambda_2\tau}-1\right)}{\lambda_1\lambda_2\left(\lambda_1+\lambda_2\right)\left(e^{\lambda_1\tau}-1\right)\left(e^{\lambda_2\tau}-1\right)},$$

$$T_- = \frac{\tau e^{(\lambda_1+\lambda_2)\tau}}{\left(e^{\lambda_1\tau}-1\right)\left(e^{\lambda_2\tau}-1\right)} - \frac{\lambda_1^2 e^{\lambda_1\tau}\left(e^{\lambda_2\tau}-1\right)+\lambda_2^2 e^{\lambda_2\tau}\left(e^{\lambda_1\tau}-1\right)+\lambda_1\lambda_2\left(e^{\lambda_1\tau}-1\right)\left(e^{\lambda_2\tau}-1\right)}{\lambda_1\lambda_2\left(\lambda_1+\lambda_2\right)\left(e^{\lambda_1\tau}-1\right)\left(e^{\lambda_2\tau}-1\right)}.$$

Stationary availability factor, average specific income and average specific expenses are:

$$K_a = \frac{\lambda_1\left(1-e^{-\lambda_2\tau}\right)}{\lambda_2(\lambda_1+\lambda_2)\tau} + \frac{\lambda_2\left(1-e^{-\lambda_1\tau}\right)}{\lambda_1(\lambda_1+\lambda_2)\tau} + \frac{\left(1-e^{-\lambda_1\tau}\right)\left(1-e^{-\lambda_2\tau}\right)}{(\lambda_1+\lambda_2)\tau},$$

$$S = \frac{(c_1+c_2)\left(\lambda_1^2\left(1-e^{-\lambda_2\tau}\right)+\lambda_2^2\left(1-e^{-\lambda_1\tau}\right)+\lambda_1\lambda_2\left(1-e^{-\lambda_1\tau}\right)\left(1-e^{-\lambda_2\tau}\right)\right)}{\lambda_1\lambda_2(\lambda_1+\lambda_2)\tau} - c_2,$$

$$C = \frac{c_2\lambda_1\lambda_2(\lambda_1+\lambda_2)\tau}{\lambda_1^2\left(1-e^{-\lambda_2\tau}\right)+\lambda_2^2\left(1-e^{-\lambda_1\tau}\right)+\lambda_1\lambda_2\left(1-e^{-\lambda_1\tau}\right)\left(1-e^{-\lambda_2\tau}\right)} - c_2. \tag{3.31}$$

TABLE 3.2 Values of $K_a(\tau)$, $S(\tau)$, $C(\tau)$ Under $\tau = 5$ h

Initial data		Calculation results		
$E\alpha_1$, h	$E\alpha_2$, h	$K_a(\tau)$	$S(\tau)$, c.u./h	$C(\tau)$, c.u./h
90	70	0.999	2.994	0.002
90	50	0.998	2.991	0.004
90	10	0.993	2.963	0.015

Example. Initial data and calculation results are given in Table 3.2 under $c_1 = 3$ c.u., $c_2 = 2$ c.u.

The graphs of functions $K_a(\tau)$, $S(\tau)$, $C(\tau)$ for exponential distributions are represented in Figures 3.7, 3.8, and 3.9, respectively.

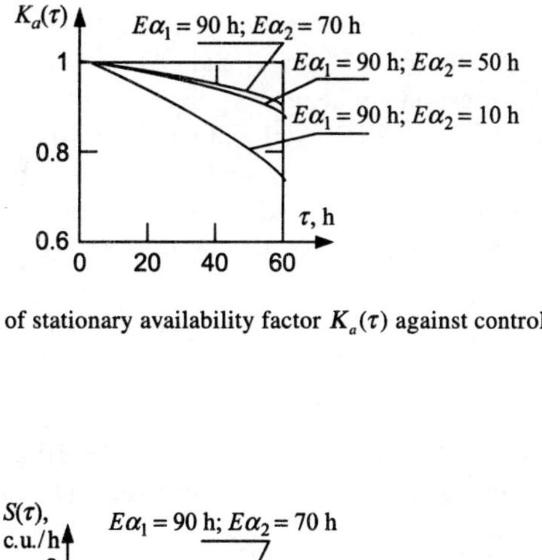

FIGURE 3.7 Graph of stationary availability factor $K_a(\tau)$ against control periodicity τ

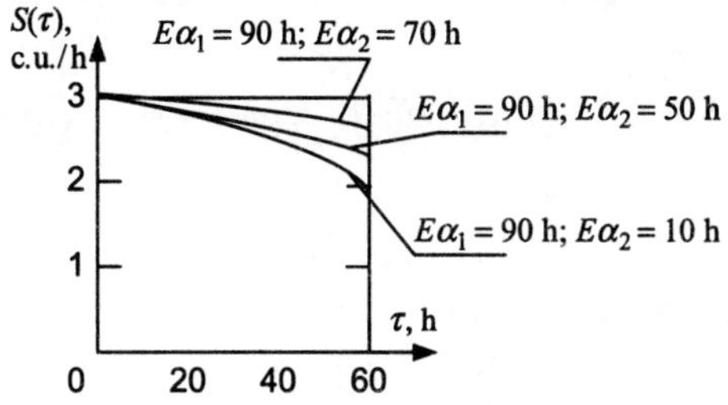

FIGURE 3.8 Graph of average specific income $S(\tau)$ against control periodicity τ

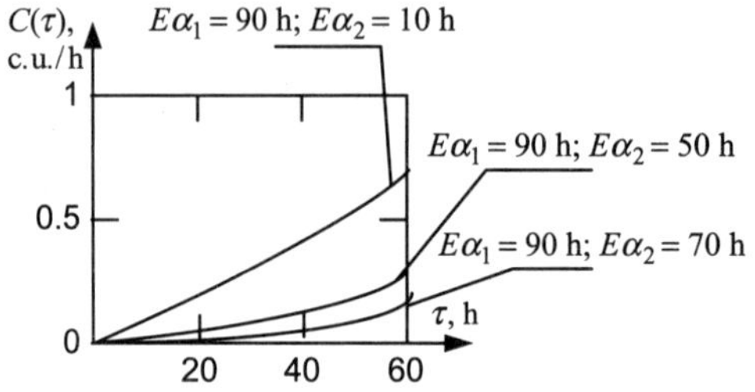

FIGURE 3.9 Graph of average specific expenses $C(\tau)$ against control periodicity τ

3.3 THE MODEL OF TWO-COMPONENT SERIAL SYSTEM WITH COMPONENTS DEACTIVATION WHILE CONTROL EXECUTION, THE DISTRIBUTION OF COMPONENTS OPERATING TF IS EXPONENTIAL

3.3.1 System Description

The system S, consisting of two serial (in reliability sense) components κ_1, κ_2 and control unit. At the initial moment, the components begin to operate, control unit is on. Components TF are RV α_1 and α_2 with DF $F_1(t)=1-e^{-\lambda_1 t}$, $F_2(t)=1-e^{-\lambda_2 t}$ and DD $f_1(t)=\lambda_1 e^{-\lambda_1 t}$, respectively. Control is carried out in random time δ with DF $R(t) = P\{\delta \leq t\}$ and DD $r(t)$. Operable components are deactivated while control execution. Failure is detected after control only. Control duration is RV γ with DF $V(t) = P\{\gamma \leq t\}$ and DD $v(t)$. After the component κ_1 failure is detected, its restoration begins, κ_2 and control are off; component κ_1 RT is RV β_1 with DF $G_1(t) = P\{\beta_1 \leq t\}$ and DD $g_1(t)$. After the component κ_2 failure is detected, its restoration begins, and control are off, component κ_2 RT is RV β_2 with DF $G_2(t) = P\{\beta_2 \leq t\}$ and DD $g_2(t)$. In case of restoration of both components, the system begins to operate after the restoration of the latter. After restoration, all the components properties are restored. RV α_1, α_2, δ, γ, β_1, and β_2 are assumed to be independent and have finite expectation.

3.3.2 Semi-Markov Model Building

We will describe the system operation by means of SMP $\xi(t)$ with a discrete-continuous phase state space. Let us introduce the following set E of semi-Markov system states:

$$E = \begin{matrix}\{3111,\ 3\hat{1}\hat{1}0,\ 1011z,\ 2101z,\ 3\hat{0}\hat{1}\hat{0},\ 3\hat{1}\hat{0}0,\ 32\hat{1}2,\ 3\hat{1}22,\ 1001z,\ 2001z \\ 3\hat{0}\hat{0}0,\ 3222,\ 1\hat{1}22x,\ 22\hat{1}2x\}.\end{matrix}$$

State codes have the following conceptual sense:

3111 – components κ_1, κ_2 have begun to operate, control is on;

$3\hat{1}\hat{1}0$ – control has begun, components κ_1, κ_2 are in up-state and are off;

$1011z$ – component κ_1 has failed, component κ_2 is in up-state, time $z > 0$ is left till the control beginning;

$2101z$ – component κ_1 has failed, component κ_2 is in up-state, time $z > 0$ is left till the control beginning;

$3\hat{0}\hat{1}0$ – control has begun, components κ_1, κ_2 are deactivated, component κ_1 is in down and component κ_2 is in up-state;

$3\hat{1}\hat{0}0$ – control has begun, components κ_1, κ_2 are deactivated, component κ_1 is in down and component κ_2 is in up-state;

$32\hat{1}2$ – control has ended, component κ_1 failure is detected, its restoration has begun, component κ_2 is in up-state and is off;

$3\hat{1}22$ – control has ended, component κ_1 failure is detected, its restoration has begun, component κ_2 is in up-state and is off;

$1001z$ – component κ_1 has failed, component κ_2 is in down-state, time $z > 0$ is left till control beginning;

$2001z$ – component κ_2 has failed, component κ_1 is in down-state, time $z > 0$ is left till control beginning;

$3\hat{0}\hat{0}0$ – control has begun, components κ_1, κ_2 are in down-state and are off;

3222 – control has ended, components κ_1, κ_2 failures are detected, their restoration has begun;

$1\hat{1}22x$ – component κ_1 has been restored and deactivated, time $x > 0$ is left till component κ_2 restoration, control is suspended;

$22\hat{1}2x$ – component κ_2 has been restored and deactivated, time $x > 0$ is left till component κ_1 restoration, control is suspended.

Time diagram of system operation and system transition graph are presented in Figures 3.10 and 3.11 correspondingly.

Let us define system sojourn times in states.

$$\begin{aligned}
&\theta_{3111} = \alpha_1 \wedge \alpha_2 \wedge \delta, \theta_{3\hat{1}\hat{1}0} = \gamma, \theta_{1011z} = \alpha_2 \wedge z, \theta_{2101z} = \alpha_1 \wedge z, \theta_{3\hat{0}\hat{1}0} = \gamma, \theta_{3\hat{1}\hat{0}0} = \gamma,\\
&\theta_{32\hat{1}2} = \beta_1, \theta_{3\hat{1}22} = \beta_2, \theta_{1001z} = z, \theta_{2001z} = z,\\
&\theta_{3\hat{0}\hat{0}0} = \gamma, \theta_{3222} = \beta_1 \wedge \beta_2, \theta_{1\hat{1}22x} = x, \theta_{22\hat{1}2x} = x.
\end{aligned} \tag{3.32}$$

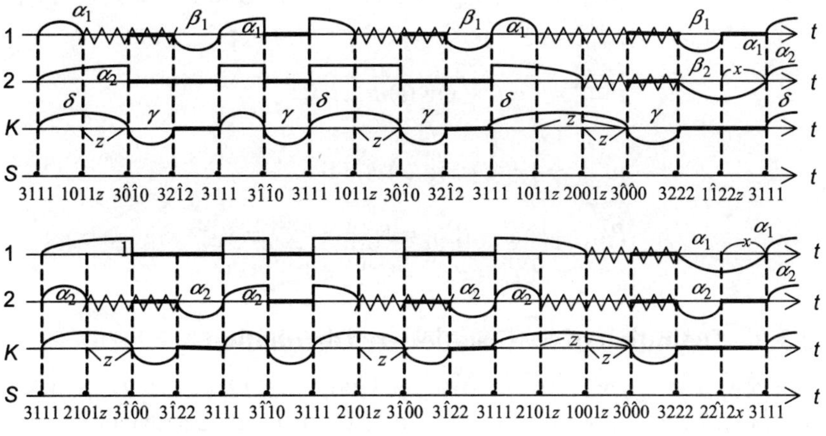

FIGURE 3.10 Time diagram of system operation

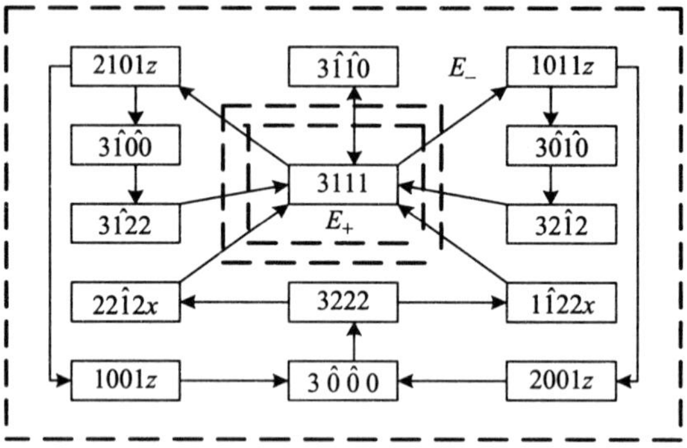

FIGURE 3.11 System transition graph for parallel components connection

---

Description of system transition events. The transition events from the states 3111, 2101*z*, 1011*z*, 3222 are defined similarly to ones described in Section 3.1. Transitions $3\hat{1}\hat{1}0 \to 3111$, $3\hat{0}\hat{1}0 \to 32\hat{1}2$, $3\hat{1}\hat{0}0 \to 3\hat{1}22$, $32\hat{1}2 \to 3111$, $3\hat{1}22 \to 3111$, $1001z \to 3\hat{0}\hat{0}0$, $2001z \to 3\hat{0}\hat{0}0$, $3\hat{0}\hat{0}0 \to 3222$, $1\hat{1}22x \to 3111$, $22\hat{1}2x \to 3111$ occur with the unity probability.

Let us find EMC $\{\xi_n;\ n \ge 0\}$ transition probabilities:

$$
\begin{aligned}
&p_{3111}^{1011z} = \lambda_1 \int_0^\infty e^{-(\lambda_1+\lambda_2)t} r(t+z)dt, z>0;\\
&p_{3111}^{2101z} = \lambda_2 \int_0^\infty e^{-(\lambda_1+\lambda_2)t} r(t+z)dt, z>0;\\
&p_{3111}^{3\hat{1}\hat{1}0} = \int_0^\infty e^{-(\lambda_1+\lambda_2)t} r(t)dt;\\
&p_{2101z}^{3\hat{1}\hat{0}0} = e^{-\lambda_1 z}; p_{2101z}^{1001z_1} = \lambda_1 e^{-\lambda_1(z-z_1)}, 0<z_1<z;\\
&p_{1011z}^{3\hat{0}\hat{1}0} = e^{-\lambda_2 z}; p_{1011z}^{2001z_1} = \lambda_2 e^{-\lambda_2(z-z_1)}, 0<z_1<z;\\
&p_{3222}^{1\hat{1}22x} = \int_0^\infty g_2(t+x)g_1(t)dt, x>0;\\
&p_{3222}^{22\hat{1}2x} = \int_0^\infty g_1(t+x)g_2(t)dt, x>0.\\
&P_{3\hat{1}\hat{1}0}^{3111} = P_{3\hat{1}\hat{0}0}^{3\hat{1}22} = P_{1001z}^{3\hat{0}\hat{0}0} = P_{3\hat{0}\hat{1}0}^{32\hat{1}2} = P_{2001z}^{3\hat{0}\hat{0}0} = P_{3\hat{1}22}^{3111}\\
&\quad = P_{3\hat{0}\hat{0}0}^{3222} = P_{32\hat{1}2}^{3111} = P_{1\hat{1}22x}^{3111} = P_{22\hat{1}2x}^{3111} = 1.
\end{aligned}
\tag{3.33}
$$

3.3.3 Definition of EMC Stationary Distribution

Denote by $\rho(3111)$, $\rho(3\hat{1}\hat{1}0)$, $\rho(3\hat{0}\hat{1}0)$, $\rho(3\hat{1}\hat{0}0)$, $\rho(32\hat{1}2)$, $\rho(3\hat{1}22)$, $\rho(3\hat{0}\hat{0}0)$, and $\rho(3222)$, the values of EMC $\{\xi_n;\ n \ge 0\}$ stationary distribution in states 3111, $3\hat{1}\hat{1}0$, $3\hat{0}\hat{1}0$, $3\hat{1}\hat{0}0$, $32\hat{1}2$, $3\hat{1}22$, $3\hat{0}\hat{0}0$, and 3222, respectively, and assume the existence of stationary densities $\rho(1011z)$, $\rho(2101z)$, $\rho(1001z)$, $\rho(2001z)$, $\rho(1\hat{1}22x)$, $\rho(22\hat{1}2x)$ for states $1011z$, $2101z$, $1001z$, $2001z$, $1\hat{1}22x$, $22\hat{1}2x$ correspondingly. Construct the system of integral equation (3.34) for them.

$$
\begin{cases}
\rho(3111) = \rho(3\hat{1}\hat{1}0) + \rho(32\hat{1}2) + \rho(3\hat{1}22) + \int_0^\infty \rho(1\hat{1}22x)\,dx + \int_0^\infty \rho(22\hat{1}2x)\,dx, \\
\rho(3\hat{1}\hat{1}0) = \rho(3111)\int_0^\infty e^{-(\lambda_1+\lambda_2)t} r(t)\,dt, \\
\rho(1011z) = \rho(3111)\lambda_1 \int_0^\infty e^{-(\lambda_1+\lambda_2)t} r(t+z)\,dt, \\
\rho(2101z) = \rho(3111)\lambda_2 \int_0^\infty e^{-(\lambda_1+\lambda_2)t} r(t+z)\,dt, \\
\rho(3\hat{0}\hat{1}0) = \int_0^\infty \rho(1011z) e^{-\lambda_2 z}\,dz, \\
\rho(3\hat{1}\hat{0}0) = \int_0^\infty \rho(2101z) e^{-\lambda_1 z}\,dz, \\
\rho(32\hat{1}2) = \rho(3\hat{0}\hat{1}0), \\
\rho(3\hat{1}22) = \rho(3\hat{1}\hat{0}0), \\
\rho(1001z) = \lambda_1 \int_0^\infty \rho(2101z+t) e^{-\lambda_1 t}\,dt, \\
\rho(2001z) = \lambda_2 \int_0^\infty \rho(1011z+t) e^{-\lambda_2 t}\,dt, \\
\rho(3\hat{0}\hat{0}0) = \int_0^\infty \rho(2001z)\,dz + \int_0^\infty \rho(1001z)\,dz, \\
\rho(3222) = \rho(3\hat{0}\hat{0}0), \\
\rho(1\hat{1}22x) = \rho(3222)\int_0^\infty g_1(t) g_2(t+x)\,dt, \\
\rho(22\hat{1}2x) = \rho(3222)\int_0^\infty g_2(t) g_1(t+x)\,dt, \\
\rho(3111) + \rho(3\hat{1}\hat{1}0) + \int_0^\infty \rho(1011z)\,dz + \int_0^\infty \rho(2101z)\,dz + \rho(3\hat{0}\hat{1}0) + \\
+\rho(3\hat{1}\hat{0}0) + \rho(32\hat{1}2) + \rho(3\hat{1}22) + \int_0^\infty \rho(1001z)\,dz + \int_0^\infty \rho(2001z)\,dz + \\
+\rho(3\hat{0}\hat{0}0) + \rho(3222) + \int_0^\infty \rho(1\hat{1}22x)\,dx + \int_0^\infty \rho(22\hat{1}2x)\,dx = 1.
\end{cases}
\tag{3.34}
$$

The last equation in the system (3.34) is a normalization requirement.

Let us introduce the notation: $\rho(3111) = \rho_0$. One can prove (Appendix D) the system of equations (3.34) has the following solution (3.35):

$$\begin{cases} \rho(3111) = \rho_0, \rho(3\hat{1}\hat{1}0) = \rho_0 \int_0^\infty e^{-(\lambda_1+\lambda_2)t} r(t)dt, \\ \rho(1011z) = \rho_0\lambda_1 \int_0^\infty e^{-(\lambda_1+\lambda_2)t} r(t+z)dt, \rho(2101z) = \rho_0\lambda_2 \int_0^\infty e^{-(\lambda_1+\lambda_2)t} r(t+z)dt, \\ \rho(3\hat{0}\hat{1}0) = \rho_0 \left(\int_0^\infty e^{-\lambda_2 t} r(t)dt - \int_0^\infty e^{-(\lambda_1+\lambda_2)t} r(t)dt \right), \\ \rho(3\hat{1}\hat{0}0) = \rho_0 \left(\int_0^\infty e^{-\lambda_1 t} r(t)dt - \int_0^\infty e^{-(\lambda_1+\lambda_2)t} r(t)dt \right), \\ \rho(32\hat{1}2) = \rho(3\hat{0}\hat{1}0), \rho(3\hat{1}22) = \rho(3\hat{1}\hat{0}0), \\ \rho(1001z) = \rho_0\lambda_1 \left(\int_0^\infty e^{-\lambda_1 t} r(t+z)dt - \int_0^\infty e^{-(\lambda_1+\lambda_2)t} r(t+z)dt \right), \\ \rho(2001z) = \rho_0\lambda_2 \left(\int_0^\infty e^{-\lambda_2 t} r(t+z)dt - \int_0^\infty e^{-(\lambda_1+\lambda_2)t} r(t+z)dt \right), \\ \rho(3\hat{0}\hat{0}0) = \rho_0 \left(1 - \int_0^\infty e^{-\lambda_1 t} r(t)dt - \int_0^\infty e^{-\lambda_2 t} r(t)dt + \int_0^\infty e^{-(\lambda_1+\lambda_2)t} r(t)dt \right), \\ \rho(3222) = \rho(3\hat{0}\hat{0}0), \rho(1\hat{1}22x) = \rho(3\hat{0}\hat{0}0) \int_0^\infty g_1(t) g_2(t+x)dt, \\ \rho(22\hat{1}2x) = \rho(3\hat{0}\hat{0}0) \int_0^\infty g_2(t) g_1(t+x)dt, \end{cases} \tag{3.35}$$

the value of ρ_0 can be obtained from the normalization requirement.

3.3.4 Stationary Characteristics Definition

Let us split the phase state space E into the following subsets:

$E_+ = \{3111\}$ – system up-state;

$$E_- = \{3\hat{1}\hat{1}0,\ 1011z,\ 2101z,\ 3\hat{0}\hat{1}0,\ 3\hat{1}\hat{0}0,\ 32\hat{1}2,\ 3\hat{1}22,\ 1001z,\ 2001z,\ 3\hat{0}\hat{0}0,\ 3222,\ 1\hat{1}22x,\ 22\hat{1}2x\}$$

-system down-state.

Applying (3.32), we determine the system average sojourn times in states:

$$\begin{aligned}&m(3111)=\int_0^\infty e^{-(\lambda_1+\lambda_2)}\overline{R}(t)\,dt, m(3\hat{1}\hat{1}0)=E\gamma, m(1011z)=\frac{1}{\lambda_2}\left(1-e^{-\lambda_2 z}\right),\\&m(2101z)=\frac{1}{\lambda_1}\left(1-e^{-\lambda_1 z}\right), m(3\hat{0}\hat{1}0)=E\gamma, m(3\hat{1}\hat{0}0)=E\gamma, m(32\hat{1}2)=E\beta_1,\\&m(3\hat{1}22)=E\beta_2, m(1001z)=z, m(2001z)=z, m(3\hat{0}\hat{0}0)=E\gamma,\\&m(3222)=\int_0^\infty \overline{G_1}(t)\overline{G_2}(t)\,dt, m(1\hat{1}22x)=x, m(22\hat{1}2x)=x.\end{aligned}\tag{3.36}$$

Average stationary operating TF T_+ and average stationary restoration time T_- are obtained by means of formulas given in Chapter 1 (1.17 and 1.18).

With regard to formulas (3.35) and (3.36), let us get the expressions from Chapter 1 (1.17 and 1.18).

$$\begin{aligned}\int_{E+} m(e)\rho(de)&=m(3111)\rho(3111)=\rho_0\int_0^\infty e^{-(\lambda_1+\lambda_2)t}\overline{R}(t)\,dt=\\&=\rho_0\frac{1}{\lambda_1+\lambda_2}\left(1-\int_0^\infty e^{-(\lambda_1+\lambda_2)t}r(t)\,dt\right).\end{aligned}\tag{3.37}$$

Next,

$$\begin{aligned}&\int_{E_-} m(e)\rho(de)=m(3\hat{1}\hat{1}0)\rho(3\hat{1}\hat{1}0)+m(3\hat{1}\hat{0}0)\rho(3\hat{1}\hat{0}0)+m(3\hat{0}\hat{1}0)\rho(3\hat{0}\hat{1}0)+\\&+m(3\hat{0}\hat{0}0)\rho(3\hat{0}\hat{0}0)+m(32\hat{1}2)\rho(32\hat{1}2)+m(3\hat{1}22)\rho(3\hat{1}22)+m(3222)\rho(3222)+\\&+\int_0^\infty m(2101z)\rho(2101z)\,dz+\int_0^\infty m(1011z)\rho(1011z)\,dz+\int_0^\infty m(1001z)\rho(1001z)\,dz+\\&+\int_0^\infty m(2001z)\rho(2001z)\,dz+\int_0^\infty m(1\hat{1}22x)\rho(1\hat{1}22x)\,dx+\int_0^\infty m(22\hat{1}2x)\rho(22\hat{1}2x)\,dx=\\&=\rho_0\left(E\gamma\int_0^\infty e^{-(\lambda_1+\lambda_2)t}r(t)\,dt+E\gamma\left(\int_0^\infty e^{-\lambda_1 t}r(t)\,dt-\int_0^\infty e^{-(\lambda_1+\lambda_2)t}r(t)\,dt\right)+\right.\\&+E\gamma\left(\int_0^\infty e^{-\lambda_2 t}r(t)\,dt-\int_0^\infty e^{-(\lambda_1+\lambda_2)t}r(t)\,dt\right)+\\&+E\gamma\left(\lambda_1\int_0^\infty e^{-\lambda_1 t}\overline{R}(t)\,dt+\lambda_2\int_0^\infty e^{-\lambda_2 t}\overline{R}(t)\,dt-(\lambda_1+\lambda_2)\int_0^\infty e^{-(\lambda_1+\lambda_2)t}\overline{R}(t)\,dt\right)+\\&+E\beta_1\left(\int_0^\infty e^{-\lambda_2 t}r(t)\,dt-\int_0^\infty e^{-(\lambda_1+\lambda_2)t}r(t)\,dt\right)+E\beta_2\left(\int_0^\infty e^{-\lambda_1 t}r(t)\,dt-\int_0^\infty e^{-(\lambda_1+\lambda_2)t}r(t)\,dt\right)+\end{aligned}$$

$$+\frac{\lambda_2}{\lambda_1}\int_0^\infty\left(1-e^{-\lambda_1 z}\right)dz\int_0^\infty e^{-(\lambda_1+\lambda_2)t}r(t+z)dt+\frac{\lambda_1}{\lambda_2}\int_0^\infty\left(1-e^{-\lambda_2 z}\right)dz\int_0^\infty e^{-(\lambda_1+\lambda_2)t}r(t+z)dt+$$
$$+\lambda_1\int_0^\infty z\,dz\left(\int_0^\infty e^{-\lambda_1 t}r(t+z)dt-\int_0^\infty e^{-(\lambda_1+\lambda_2)t}r(t+z)dt\right)+$$
$$+\lambda_2\int_0^\infty z\,dz\left(\int_0^\infty e^{-\lambda_2 t}r(t+z)dt-\int_0^\infty e^{-(\lambda_1+\lambda_2)t}r(t+z)dt\right)\Bigg)+\rho(3\hat{0}\hat{0}0)E(\beta_1\wedge\beta_2)+ \tag{3.38}$$
$$+\rho(3\hat{0}\hat{0}0)\int_0^\infty x\,dx\int_0^\infty g_2(t+x)g_1(t)dt+\rho(3\hat{0}\hat{0}0)\int_0^\infty x\,dx\int_0^\infty g_1(t+x)g_2(t)dt.$$

By making some transformations, one can show the expression (3.38) takes the form:

$$\int_{E_-} m(e)\rho(de)=\rho_0\Bigg(E\gamma+E\delta-\int_0^\infty e^{-(\lambda_1+\lambda_2)t}\bar{R}(t)dt+$$
$$+E\beta_1\left(1-\int_0^\infty e^{-\lambda_1 t}r(t)dt\right)+E\beta_2\left(1-\int_0^\infty e^{-\lambda_2 t}r(t)dt\right)- \tag{3.39}$$
$$-E(\beta_1\wedge\beta_2)\left(1-\int_0^\infty e^{-\lambda_1 t}r(t)dt-\int_0^\infty e^{-\lambda_2 t}r(t)dt+\int_0^\infty e^{-(\lambda_1+\lambda_2)t}r(t)dt\right)\Bigg).$$

Next,

$$\int_{E_+} P(e,E_-)\rho(de)=\rho_0\Bigg(\lambda_1\int_0^\infty dz\int_0^\infty e^{-(\lambda_1+\lambda_2)t}r(t+z)dt+$$
$$+\lambda_2\int_0^\infty dz\int_0^\infty e^{-(\lambda_1+\lambda_2)t}r(t+z)dt+\int_0^\infty e^{-(\lambda_1+\lambda_2)t}r(t)dt\Bigg)=$$
$$=\rho_0\Bigg(\lambda_1\int_0^\infty e^{-(\lambda_1+\lambda_2)t}dt\int_0^\infty r(t+z)dz+$$
$$+\lambda_2\int_0^\infty e^{-(\lambda_1+\lambda_2)t}dt\int_0^\infty r(t+z)dz+1-(\lambda_1+\lambda_2)\int_0^\infty e^{-(\lambda_1+\lambda_2)t}\bar{R}(t)dt\Bigg)= \tag{3.40}$$
$$=\rho_0\Bigg(\lambda_1\int_0^\infty e^{-(\lambda_1+\lambda_2)t}\bar{R}(t)dt+$$
$$+\lambda_2\int_0^\infty e^{-(\lambda_1+\lambda_2)t}\bar{R}(t)dt+1-(\lambda_1+\lambda_2)\int_0^\infty e^{-(\lambda_1+\lambda_2)t}\bar{R}(t)dt\Bigg)=\rho_0.$$

Thus, taking into account formulas (3.37), (3.39), and (3.40) we obtain average stationary operating TF T_+ and average stationary restoration time T_-:

$$T_+ = \frac{1}{\lambda_1+\lambda_2}\left(1-\int_0^\infty e^{-(\lambda_1+\lambda_2)t} r(t)\,dt\right), \tag{3.41}$$

$$\begin{aligned} T_- = {} & \left(E\gamma + E\delta - \int_0^\infty e^{-(\lambda_1+\lambda_2)t}\bar{R}(t)\,dt + E\beta_1\left(1-\int_0^\infty e^{-\lambda_1 t} r(t)\,dt\right)+\right. \\ & + E\beta_2\left(1-\int_0^\infty e^{-\lambda_2 t} r(t)\,dt\right) - E(\beta_1 \wedge \beta_2)\left(1-\int_0^\infty e^{-\lambda_1 t} r(t)\,dt -\right. \\ & \left.\left. - \int_0^\infty e^{-\lambda_2 t} r(t)\,dt + \int_0^\infty e^{-(\lambda_1+\lambda_2)t} r(t)\,dt\right)\right). \end{aligned} \tag{3.42}$$

In correspondence with formulas (3.41) and (3.42), stationary availability factor is:

$$\begin{aligned} K_a = {} & \frac{1}{\lambda_1+\lambda_2}\left(1-\int_0^\infty e^{-(\lambda_1+\lambda_2)t} r(t)\,dt\right) \Bigg/ \left(E\gamma + E\delta + E\beta_1\left(1-\int_0^\infty e^{-\lambda_1 t} r(t)\,dt\right)+\right. \\ & + E\beta_2\left(1-\int_0^\infty e^{-\lambda_2 t} r(t)\,dt\right) - E(\beta_1 \wedge \beta_2)\left(1-\int_0^\infty e^{-\lambda_1 t} r(t)\,dt -\right. \\ & \left.\left. - \int_0^\infty e^{-\lambda_2 t} r(t)\,dt + \int_0^\infty e^{-(\lambda_1+\lambda_2)t} r(t)\,dt\right)\right). \end{aligned} \tag{3.43}$$

Now we can find system efficiency characteristics: average specific income S per calendar time unit and average specific expenses C per time unit of up-state with the help of Chapter 1 (1.21 and 1.22).

For the given system, the functions $f_s(e), f_c(e)$ are as follows:

$$\begin{aligned} f_s(e) &= \begin{cases} c_1, & e = 3111, \\ -c_2, & e \in \{32\hat{1}2,\ 3\hat{1}22,\ 3222,\ 1\hat{1}22x,\ 22\hat{1}2x\}, \\ -c_3, & e \in \{3\hat{1}\hat{1}0,\ 3\hat{0}\hat{1}0,\ 3\hat{1}\hat{0}0,\ 3\hat{0}\hat{0}0\}, \\ -c_4, & e \in \{1011z,\ 2101z,\ 1001z,\ 2001z\}, \end{cases} \\ f_c(e) &= \begin{cases} 0, & e = 3111, \\ c_2, & e \in \{32\hat{1}2,\ 3\hat{1}22,\ 3222,\ 1\hat{1}22x,\ 22\hat{1}2x\}, \\ c_3, & e \in \{3\hat{1}\hat{1}0,\ 30\hat{1}0,\ 3\hat{1}\hat{0}0,\ 3\hat{0}\hat{0}0\}, \\ c_4, & e \in \{1011z,\ 2101z,\ 1001z,\ 2001z\}. \end{cases} \end{aligned} \tag{3.44}$$

Here c_1 is the income per time unit of components operation, c_2 are expenses per time unit of components restoration, c_3 are expenses per time unit of control, and c_4 are expenses per time unit of latent failure.

By using formulas (3.35), (3.36), and (3.44), we get the expressions included into formulas given in Chapter 1 (1.22 and 1.23):

$$\int_E m(e)f_s(e)\rho(de)=c_1\rho(3111)m(3111)-c_2\Bigg(m(32\hat{1}2)\rho(32\hat{1}2)+m(3\hat{1}22)\rho(3\hat{1}22)+$$
$$+m(3222)\rho(3222)+\int_0^\infty m(1\hat{1}22x)\rho(1\hat{1}22x)dx+\int_0^\infty m(22\hat{1}2x)\rho(22\hat{1}2x)dx\Bigg)-$$
$$-c_3\Big(m(3\hat{1}\hat{1}0)\rho(3\hat{1}\hat{1}0)+m(3\hat{0}\hat{1}0)\rho(3\hat{0}\hat{1}0)+m(3\hat{1}\hat{0}0)\rho(3\hat{1}\hat{0}0)+m(3\hat{0}\hat{0}0)\rho(3\hat{0}\hat{0}0)\Big)-$$
$$-c_4\Bigg(\int_0^\infty m(1011z)\rho(1011z)dz+\int_0^\infty m(2101z)\rho(2101z)dz+$$
$$+\int_0^\infty m(1001z)\rho(1001z)dz+\int_0^\infty m(2001z)\rho(2001z)dz\Bigg)=$$
$$=\rho_0\Bigg((c_1+c_4)\frac{1}{\lambda_1+\lambda_2}\Bigg(1-\int_0^\infty e^{-(\lambda_1+\lambda_2)t}r(t)dt\Bigg)-c_3E\gamma-c_4E\delta-$$
$$-c_2\Bigg(E\beta_1\Bigg(1-\int_0^\infty e^{-\lambda_1 t}r(t)dt\Bigg)+E\beta_2\Bigg(1-\int_0^\infty e^{-\lambda_2 t}r(t)dt\Bigg)-$$
$$-E(\beta_1\wedge\beta_2)\Bigg(1-\int_0^\infty e^{-\lambda_1 t}r(t)dt-\int_0^\infty e^{-\lambda_2 t}r(t)dt+\int_0^\infty e^{-(\lambda_1+\lambda_2)t}r(t)dt\Bigg)\Bigg)\Bigg). \quad (3.45)$$

Next,

$$\int_E m(e)f_c(e)\rho(de)=c_3\Bigg(m(3\hat{1}\hat{1}0)\rho(3\hat{1}\hat{1}0)+m(3\hat{0}\hat{1}0)\rho(3\hat{0}\hat{1}0)+m(3\hat{1}\hat{0}0)\rho(3\hat{1}\hat{0}0)+$$
$$+m(3\hat{0}\hat{0}0)\rho(3\hat{0}\hat{0}0)\Bigg)+c_2\Bigg(m(32\hat{1}2)\rho(32\hat{1}2)+m(3\hat{1}22)\rho(3\hat{1}22)+$$
$$+m(3222)\rho(3222)+\int_0^\infty m(1\hat{1}22x)\rho(1\hat{1}22x)dx+\int_0^\infty m(22\hat{1}2x)\rho(22\hat{1}2x)dx\Bigg)+$$
$$+c_4\Bigg(\int_0^\infty m(1011z)\rho(1011z)dz+\int_0^\infty m(2001z)\rho(2001z)dz+$$
$$+\int_0^\infty m(2101z)\rho(2101z)dz+\int_0^\infty m(1001z)\rho(1001z)dz\Bigg)= \quad (3.46)$$
$$=\rho_0\Bigg(c_3E\gamma+c_4\Bigg(E\delta-\frac{1}{\lambda_1+\lambda_2}\Bigg(1-\int_0^\infty e^{-(\lambda_1+\lambda_2)t}r(t)dt\Bigg)\Bigg)+$$
$$+c_2\Bigg(E\beta_1\Bigg(1-\int_0^\infty e^{-\lambda_1 t}r(t)dt\Bigg)+E\beta_2\Bigg(1-\int_0^\infty e^{-\lambda_2 t}r(t)dt\Bigg)-$$
$$-E(\beta_1\wedge\beta_2)\Bigg(1-\int_0^\infty e^{-\lambda_1 t}r(t)dt-\int_0^\infty e^{-\lambda_2 t}r(t)dt+\int_0^\infty e^{-(\lambda_1+\lambda_2)t}r(t)dt\Bigg)\Bigg)\Bigg).$$

With regard to (3.37), (3.39), (3.45), and (3.46), average specific income and average specific expenses are:

$$S=\left((c_1+c_4)\frac{1}{\lambda_1+\lambda_2}\left(1-\int_0^\infty e^{-(\lambda_1+\lambda_2)t}r(t)dt\right)-c_3E\gamma-c_4E\delta-\right.$$
$$-c_2\left(E\beta_1\left(1-\int_0^\infty e^{-\lambda_1 t}r(t)dt\right)+E\beta_2\left(1-\int_0^\infty e^{-\lambda_2 t}r(t)dt\right)-\right.$$
$$\left.\left.-E(\beta_1\wedge\beta_2)\left(1-\int_0^\infty e^{-\lambda_1 t}r(t)dt-\int_0^\infty e^{-\lambda_2 t}r(t)dt+\int_0^\infty e^{-(\lambda_1+\lambda_2)t}r(t)dt\right)\right)\right)\Big/$$
$$\left(E\gamma+E\delta+E\beta_1\left(1-\int_0^\infty e^{-\lambda_1 t}r(t)dt\right)+E\beta_2\left(1-\int_0^\infty e^{-\lambda_2 t}r(t)dt\right)-\right.$$
$$\left.-E(\beta_1\wedge\beta_2)\left(1-\int_0^\infty e^{-\lambda_1 t}r(t)dt-\int_0^\infty e^{-\lambda_2 t}r(t)dt+\int_0^\infty e^{-(\lambda_1+\lambda_2)t}r(t)dt\right)\right), \tag{3.47}$$

$$C=\left(c_3E\gamma+c_4\left(E\delta-\frac{1}{\lambda_1+\lambda_2}\left(1-\int_0^\infty e^{-(\lambda_1+\lambda_2)t}r(t)dt\right)\right)+\right.$$
$$+c_2\left(E\beta_1\left(1-\int_0^\infty e^{-\lambda_1 t}r(t)dt\right)+EM\beta_2\left(1-\int_0^\infty e^{-\lambda_2 t}r(t)dt\right)-\right.$$
$$\left.\left.-E(\beta_1\wedge\beta_2)\left(1-\int_0^\infty e^{-\lambda_1 t}r(t)dt-\int_0^\infty e^{-\lambda_2 t}r(t)dt+\int_0^\infty e^{-(\lambda_1+\lambda_2)t}r(t)dt\right)\right)\right)\Big/$$
$$\frac{1}{\lambda_1+\lambda_2}\left(1-\int_0^\infty e^{-(\lambda_1+\lambda_2)t}r(t)dt\right). \tag{3.48}$$

Denoting by $\tilde{r}(\lambda)=\int_0^\infty e^{-\lambda t}r(t)dt$ Laplace transform of the function $r(t)$, we have:

$$\tilde{r}(\lambda_1)=\int_0^\infty e^{-\lambda_1 t}r(t)dt;\tilde{r}(\lambda_2)=\int_0^\infty e^{-\lambda_2 t}r(t)dt;\tilde{r}(\lambda_1+\lambda_2)=$$
$$=\int_0^\infty e^{-(\lambda_1+\lambda_2)t}r(t)dt.$$

Then formulas (3.41), (3.42), (3.43), (3.47), and (3.48) can be represented as:

$$T_+ = \frac{1-\tilde{r}(\lambda_1+\lambda_2)}{\lambda_1+\lambda_2},$$

$$T_- = E\gamma + E\delta - \frac{1-\tilde{r}(\lambda_1+\lambda_2)}{\lambda_1+\lambda_2} + E\beta_1(1-\tilde{r}(\lambda_1)) + E\beta_2(1-\tilde{r}(\lambda_2)) -$$
$$-E(\beta_1 \wedge \beta_2)(1-\tilde{r}(\lambda_1)-\tilde{r}(\lambda_2)+\tilde{r}(\lambda_1+\lambda_2)),$$

$$K_a = \frac{1}{\lambda_1+\lambda_2}(1-\tilde{r}(\lambda_1+\lambda_2)) \Big/ \big(E\gamma + E\delta + E\beta_1(1-\tilde{r}(\lambda_1)) +$$
$$+E\beta_2(1-\tilde{r}(\lambda_2)) - E(\beta_1 \wedge \beta_2)(1-\tilde{r}(\lambda_1)-\tilde{r}(\lambda_2)+\tilde{r}(\lambda_1+\lambda_2))\big),$$

$$S = \left(\frac{(c_1+c_4)}{\lambda_1+\lambda_2}(1-\tilde{r}(\lambda_1+\lambda_2)) - c_3E\gamma - c_4E\delta - c_2\big(E\beta_1(1-\tilde{r}(\lambda_1)) +\right.$$
$$\left.+E\beta_2(1-\tilde{r}(\lambda_2)) - E(\beta_1 \wedge \beta_2)(1-\tilde{r}(\lambda_1)-\tilde{r}(\lambda_2)+\tilde{r}(\lambda_1+\lambda_2))\big)\right) \Big/ \big(E\gamma + E\delta +$$
$$+E\beta_1(1-\tilde{r}(\lambda_1)) + E\beta_2(1-\tilde{r}(\lambda_2)) - E(\beta_1 \wedge \beta_2)(1-\tilde{r}(\lambda_1)-\tilde{r}(\lambda_2)+\tilde{r}(\lambda_1+\lambda_2))\big),$$

$$C = \left(c_3E\gamma + c_4\left(E\delta - \frac{1-\tilde{r}(\lambda_1+\lambda_2)}{\lambda_1+\lambda_2}\right) + c_2\big(E\beta_1(1-\tilde{r}(\lambda_1)) + E\beta_2(1-\tilde{r}(\lambda_2)) -\right.$$
$$\left.-E(\beta_1 \wedge \beta_2)(1-\tilde{r}(\lambda_1)-\tilde{r}(\lambda_2)+\tilde{r}(\lambda_1+\lambda_2))\big)\right) \Big/ \frac{1-\tilde{r}(\lambda_1+\lambda_2)}{\lambda_1+\lambda_2}.$$

Consider the case of nonrandom control periodicity $\tau > 0$: $R(t) = 1(t - \tau)$, $E\delta = \tau$.

Average stationary operating TF T_+ is:

$$T_+ = \frac{1-e^{-(\lambda_1+\lambda_2)\tau}}{\lambda_1+\lambda_2} = \frac{E\alpha_1 E\alpha_2}{E\alpha_1 + E\alpha_2}\left(1-e^{-(\lambda_1+\lambda_2)\tau}\right).$$

Average stationary restoration time T_- is:

$$T_- = E\gamma + E\delta - \frac{E\alpha_1 E\alpha_2}{E\alpha_1 + E\alpha_2}\left(1-e^{-(\lambda_1+\lambda_2)\tau}\right) +$$
$$+E\beta_1\left(1-e^{-\lambda_1\tau}\right) + E\beta_2\left(1-e^{-\lambda_2\tau}\right) - E(\beta_1 \wedge \beta_2)(1-e^{-\lambda_1\tau})(1-e^{-\lambda_2\tau}).$$

The following equality is true for the stationary availability factor:

$$K_a = \frac{E\alpha_1 E\alpha_2}{E\alpha_1 + E\alpha_2}\left(1-e^{-(\lambda_1+\lambda_2)\tau}\right) \Big/ \big(E\gamma + E\delta + E\beta_1\left(1-e^{-\lambda_1\tau}\right) +$$
$$+E\beta_2\left(1-e^{-\lambda_2\tau}\right) - E(\beta_1 \wedge \beta_2)\left(1-e^{-\lambda_1\tau}\right)\left(1-e^{-\lambda_2\tau}\right)\big). \tag{3.49}$$

Average specific income and average specific expenses are given by:

$$S=\left((c_1+c_4)\frac{E\alpha_1 E\alpha_2}{E\alpha_1+E\alpha_2}\left(1-e^{-(\lambda_1+\lambda_2)\tau}\right)-c_3E\gamma-c_4E\delta-\right.$$
$$\left.-c_2\left(E\beta_1\left(1-e^{-\lambda_1\tau}\right)+E\beta_2\left(1-e^{-\lambda_2\tau}\right)+E(\beta_1\wedge\beta_2)\left(e^{-\lambda_1\tau}-1\right)\left(1-e^{-\lambda_2\tau}\right)\right)\right)\Big/$$
$$\left(E\gamma+E\delta+E\beta_1\left(1-e^{-\lambda_1\tau}\right)+E\beta_2\left(1-e^{-\lambda_2\tau}\right)+E(\beta_1\wedge\beta_2)\left(e^{-\lambda_1 t}-1\right)\left(1-e^{-\lambda_2 t}\right)\right), \quad (3.50)$$

$$C=\left(c_3E\gamma+c_4\left(E\delta-\frac{E\alpha_1 E\alpha_2}{E\alpha_1+E\alpha_2}\left(1-e^{-(\lambda_1+\lambda_2)\tau}\right)\right)+c_2\left(E\beta_1\left(1-e^{-\lambda_1\tau}\right)+\right.\right.$$
$$\left.\left.+E\beta_2\left(1-e^{-\lambda_2\tau}\right)-E(\beta_1\wedge\beta_2)\left(1-e^{-\lambda_1\tau}\right)\left(1-e^{-\lambda_2\tau}\right)\right)\right)\Big/$$
$$\frac{E\alpha_1 E\alpha_2}{E\alpha_1+E\alpha_2}\left(1-e^{-(\lambda_1+\lambda_2)\tau}\right). \quad (3.51)$$

Example. Initial data and calculation results are given in Table 3.3. Average RT: $E\beta_1 = 0.100$ h, $E\beta_2 = 0.066$ h control duration is $E\gamma = 0.125$ h, $c_1 = 5$ c.u., $c_2 = 4$ c.u., $c_3 = 3$ c.u., $c_4 = 2$ c.u.

TABLE 3 .3 Values of $K_a(\tau)$, $S(\tau)$, $C(\tau)$ Under $\tau = 5$ h

Initial data		Calculation results		
$E\alpha_1$, h	$E\alpha_2$, h	$K_a(\tau)$	$S(\tau)$, c.u./h	$C(\tau)$, c.u./h
90	70	0.9147	4.352	0.242
90	50	0.9018	4.262	0.274
90	10	0.7455	3.164	0.756

3.4 THE MODEL OF TWO-COMPONENT PARALLEL SYSTEM WITH COMPONENTS DEACTIVATION WHILE CONTROL EXECUTION, THE DISTRIBUTION OF COMPONENTS OPERATING TF IS EXPONENTIAL

In the present Section, stationary characteristics of the system investigated in Section 3.3 in case of components κ_1 and κ_2 parallel connection (in reliability sense). System transition graph, under the condition of components κ_1 and κ_2 parallel connection, is given in Figure 3.12.

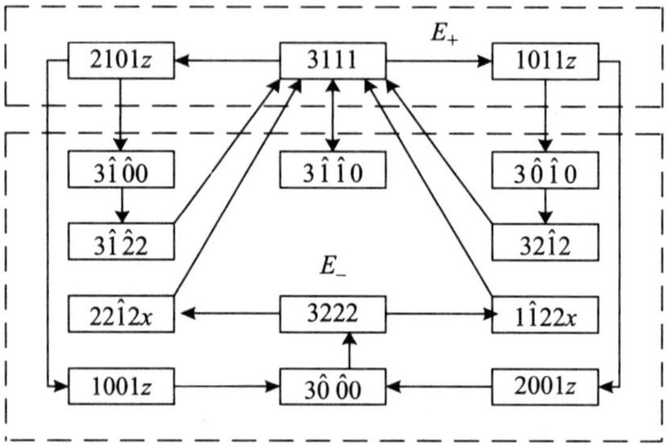

FIGURE 3.12 System transition graph for components parallel connection

3.4.1 Definition of EMC Stationary Distribution

In the case, system phase state space E can be split into two subsets:

$E_+ = \{3111, 2101z, 1011z\}$ – system up-state;

$E_- = \{3\hat{1}\hat{1}0, 3\hat{0}\hat{1}0, 3\hat{1}\hat{0}0, 32\hat{1}2, 3\hat{1}22, 3\hat{0}\hat{0}0, 3222, 1001z, 2001z, 1\hat{1}22x, 22\hat{1}2x\}$ – system down-state.

Average stationary operating TF T_+ and average stationary restoration time T_- can be obtained by means of Chapter 1 (1.17 and 1.18), with regard to (3.32), (3.33), and (3.35).

$$\int_{E_+} m(e)\rho(de) = \rho_0\left(\int_0^\infty e^{-\lambda_1 z}\bar{R}(z)dz + \int_0^\infty e^{-\lambda_2 z}\bar{R}(z)dz - \int_0^\infty e^{-(\lambda_1+\lambda_2)z}\bar{R}(z)dz\right) =$$

$$= \rho_0\left(\frac{1}{\lambda_1}\left(1-\int_0^\infty e^{-\lambda_1 t}r(t)dt\right) + \frac{1}{\lambda_2}\left(1-\int_0^\infty e^{-\lambda_2 t}r(t)dt\right) - \right. \tag{3.52}$$

$$\left. - \frac{1}{\lambda_1+\lambda_2}\left(1-\int_0^\infty e^{-(\lambda_1+\lambda_2)t}r(t)dt\right)\right).$$

Next,

$$\int_{E_-} m(e)\rho(de) = \rho_0\left(E\gamma + E\delta - \frac{1}{\lambda_1}\left(1-\int_0^\infty e^{-\lambda_1 t}r(t)dt\right) - \right.$$

$$-\frac{1}{\lambda_2}\left(1-\int_0^\infty e^{-\lambda_2 t}r(t)dt\right) + \frac{1}{\lambda_1+\lambda_2}\left(1-\int_0^\infty e^{-(\lambda_1+\lambda_2)t}r(t)dt\right) +$$

$$+ E\beta_1\left(1-\int_0^\infty e^{-\lambda_1 t}r(t)dt\right) + E\beta_2\left(1-\int_0^\infty e^{-\lambda_2 t}r(t)dt\right) -$$

$$\left. - E(\beta_1 \wedge \beta_2)\left(1-\int_0^\infty e^{-\lambda_1 y}r(y)dy - \int_0^\infty e^{-\lambda_2 y}r(y)dy + \int_0^\infty e^{-(\lambda_1+\lambda_2)y}r(y)dy\right)\right), \tag{3.53}$$

besides,

$$\int_{E_+} P(e,E_-)\rho(de) = \rho_0\left(1-(\lambda_1+\lambda_2)\int_0^\infty e^{-(\lambda_1+\lambda_2)t}\overline{R}(t)\,dt + \right.$$
$$\left. + \lambda_1\int_0^\infty e^{-(\lambda_1+\lambda_2)t}\overline{R}(t)\,dt + \lambda_2\int_0^\infty e^{-(\lambda_1+\lambda_2)t}\overline{R}(t)\,dt\right) = \rho_0. \qquad (3.54)$$

In such a way, applying (3.52), (3.53), and (3.54), we get the expressions for average stationary operating TF T_+ and average stationary restoration time T_-:

$$T_+ = \frac{1}{\lambda_1}\left(1-\int_0^\infty e^{-\lambda_1 t} r(t)\,dt\right) + \frac{1}{\lambda_2}\left(1-\int_0^\infty e^{-\lambda_2 t} r(t)\,dt\right) -$$
$$- \frac{1}{\lambda_1+\lambda_2}\left(1-\int_0^\infty e^{-(\lambda_1+\lambda_2)t} r(t)\,dt\right), \qquad (3.55)$$

$$T_- = E\gamma + E\delta - \frac{1}{\lambda_1}\left(1-\int_0^\infty e^{-\lambda_1 t} r(t)\,dt\right) - \frac{1}{\lambda_2}\left(1-\int_0^\infty e^{-\lambda_2 t} r(t)\,dt\right) +$$
$$+ \frac{1}{\lambda_1+\lambda_2}\left(1-\int_0^\infty e^{-(\lambda_1+\lambda_2)t} r(t)\,dt\right) + E\beta_1\left(1-\int_0^\infty e^{-\lambda_1 t} r(t)\,dt\right) +$$
$$+ E\beta_2\left(1-\int_0^\infty e^{-\lambda_2 t} r(t)\,dt\right) - E(\beta_1 \wedge \beta_2)\left(1-\int_0^\infty e^{-\lambda_1 t} r(t)\,dt - \right.$$
$$\left. - \int_0^\infty e^{-\lambda_2 t} r(t)\,dt + \int_0^\infty e^{-(\lambda_1+\lambda_2)t} r(t)\,dt\right). \qquad (3.56)$$

With regard to formulas (3.55) and (3.56), stationary availability factor is:

$$K_a = \left(\frac{1}{\lambda_1}\left(1-\int_0^\infty e^{-\lambda_1 t} r(t)\,dt\right) + \frac{1}{\lambda_2}\left(1-\int_0^\infty e^{-\lambda_2 t} r(t)\,dt\right) - \right.$$
$$\left. - \frac{1}{\lambda_1+\lambda_2}\left(1-\int_0^\infty e^{-(\lambda_1+\lambda_2)t} r(t)\,dt\right)\right) \Bigg/ \left(E\gamma + E\delta + E\beta_1\left(1-\int_0^\infty e^{-\lambda_1 t} r(t)\,dt\right) + \right.$$
$$+ E\beta_2\left(1-\int_0^\infty e^{-\lambda_2 t} r(t)\,dt\right) - E(\beta_1 \wedge \beta_2)\left(1-\int_0^\infty e^{-\lambda_1 t} r(t)\,dt - \right.$$
$$\left.\left. - \int_0^\infty e^{-\lambda_2 t} r(t)\,dt + \int_0^\infty e^{-(\lambda_1+\lambda_2)t} r(t)\,dt\right)\right). \qquad (3.57)$$

Let us define system stationary efficiency characteristics: average specific income S per calendar time unit and average specific expenses C per time unit of up-state according to Chapter 1 (1.22 and 1.23).

Here c_1 is the income per time unit of the component up-state, c_2 denotes expenses per time unit of restoration, c_3 are expenses per time unit of control, c_4 are wastes per time unit of latent failure of the component.

For the given system, functions $f_s(e), f_c(e)$ are as follows:

$$f_s(e)=\begin{cases} c_1, & e\in\{3111,\ 2101z,\ 1011z\},\\ -c_2, & e\in\{32\hat{1}2,\ 3\hat{1}22,\ 3222,\ 1\hat{1}22x,\ 22\hat{1}2x\},\\ -c_3, & e\in\{3\hat{1}\hat{1}0,\ 3\hat{0}\hat{1}0,\ 3\hat{1}\hat{0}0,\ 3\hat{0}\hat{0}0\},\\ -c_4, & e\in\{2001z,\ 1001z\},\end{cases}$$
$$f_c(e)=\begin{cases} 0, & e\in\{3111,\ 2101z,\ 1011z\},\\ c_2, & e\in\{32\hat{1}2,\ 3\hat{1}22,\ 3222,\ 1\hat{1}22x,\ 22\hat{1}2x\},\\ c_2, & e\in\{3\hat{1}\hat{1}0,\ 3\hat{0}\hat{1}0,\ 3\hat{1}\hat{0}0,\ 3\hat{0}\hat{0}0\},\\ c_4, & e\in\{2001z,\ 1001z\}.\end{cases} \quad (3.58)$$

With regard to (3.34), (3.35), and (3.58), average specific income and average specific expenses are:

$$S=\left((c_1+c_4)\left(\frac{1}{\lambda_1}\left(1-\int_0^\infty e^{-\lambda_1 t}r(t)dt\right)+\frac{1}{\lambda_2}\left(1-\int_0^\infty e^{-\lambda_2 t}r(t)dt\right)-\right.\right.$$
$$\left.\left.-\frac{1}{\lambda_1+\lambda_2}\left(1-\int_0^\infty e^{-(\lambda_1+\lambda_2)t}r(t)dt\right)\right)-c_3E\gamma-c_4E\delta-\right.$$
$$-c_2\left(E\beta_1\left(1-\int_0^\infty e^{-\lambda_1 t}r(t)dt\right)+E\beta_2\left(1-\int_0^\infty e^{-\lambda_2 t}r(t)dt\right)-\right.$$
$$\left.\left.-E(\beta_1\wedge\beta_2)\left(1-\int_0^\infty e^{-\lambda_1 t}r(t)dt-\int_0^\infty e^{-\lambda_2 t}r(t)dt+\int_0^\infty e^{-(\lambda_1+\lambda_2)t}r(t)dt\right)\right)\right)\Big/$$
$$\left(E\gamma+E\delta+E\beta_1\left(1-\int_0^\infty e^{-\lambda_1 t}r(t)dt\right)+E\beta_2\left(1-\int_0^\infty e^{-\lambda_2 t}r(t)dt\right)-\right.$$
$$\left.-E(\beta_1\wedge\beta_2)\left(1-\int_0^\infty e^{-\lambda_1 t}r(t)dt-\int_0^\infty e^{-\lambda_2 t}r(t)dt+\int_0^\infty e^{-(\lambda_1+\lambda_2)t}r(t)dt\right)\right), \quad (3.59)$$

$$C=\left(c_3E\gamma+c_4\left(E\delta+\frac{1}{\lambda_1+\lambda_2}\left(1-\int_0^\infty e^{-(\lambda_1+\lambda_2)t}r(t)\,dt\right)-\right.\right.$$

$$\left.-\frac{1}{\lambda_1}\left(1-\int_0^\infty e^{-\lambda_1 t}r(t)\,dt\right)-\frac{1}{\lambda_2}\left(1-\int_0^\infty e^{-\lambda_2 t}r(t)\,dt\right)\right)+c_2\left(E\beta_1\left(1-\int_0^\infty e^{-\lambda_1 t}r(t)\,dt\right)+\right.$$

$$+E\beta_2\left(1-\int_0^\infty e^{-\lambda_2 t}r(t)\,dt\right)-E(\beta_1\wedge\beta_2)\left(1-\int_0^\infty e^{-\lambda_1 t}r(t)\,dt-\int_0^\infty e^{-\lambda_2 t}r(t)\,dt+\right.$$

$$\left.\left.\left.+\int_0^\infty e^{-(\lambda_1+\lambda_2)t}r(t)\,dt\right)\right)\right)\Bigg/\left(\frac{1}{\lambda_1}\left(1-\int_0^\infty e^{-\lambda_1 t}r(t)\,dt\right)+\frac{1}{\lambda_2}\left(1-\int_0^\infty e^{-\lambda_2 t}r(t)\,dt\right)-\right.$$

$$\left.-\frac{1}{\lambda_1+\lambda_2}\left(1-\int_0^\infty e^{-(\lambda_1+\lambda_2)t}r(t)\,dt\right)\right).\tag{3.60}$$

Consider the case of nonrandom control periodicity $\tau>0$: $R(t)=1(t-\tau)$, $\tau=\text{const}$, $E\delta=\tau$. Then the formulas (3.55), (3.56), (3.57), (3.59), and (3.60) can be rewritten in the following form.

Average stationary operating TF T_+ and average stationary restoration time T_- are defined by the following formulas:

$$T_+=\frac{\lambda_2^2\left(1-e^{-\lambda_1\tau}\right)+\lambda_1^2\left(1-e^{-\lambda_2\tau}\right)+\lambda_1\lambda_2\left(1-e^{-\lambda_1\tau}\right)\left(1-e^{-\lambda_2\tau}\right)}{\lambda_1\lambda_2(\lambda_1+\lambda_2)},$$

$$T_-=E\gamma+\tau-\frac{\lambda_2^2\left(1-e^{-\lambda_1\tau}\right)+\lambda_1^2\left(1-e^{-\lambda_2\tau}\right)+\lambda_1\lambda_2\left(1-e^{-\lambda_1\tau}\right)\left(1-e^{-\lambda_2\tau}\right)}{\lambda_1\lambda_2(\lambda_1+\lambda_2)}+$$

$$+E\beta_1\left(1-e^{-\lambda_1\tau}\right)+E\beta_2\left(1-e^{-\lambda_2\tau}\right)-E(\beta_1\wedge\beta_2)(1-e^{-\lambda_1\tau})(1-e^{-\lambda_2\tau}).$$

Stationary availability factor is:

$$K_a=\frac{\lambda_2^2\left(1-e^{-\lambda_1\tau}\right)+\lambda_1^2\left(1-e^{-\lambda_2\tau}\right)+\lambda_1\lambda_2\left(1-e^{-\lambda_1\tau}\right)\left(1-e^{-\lambda_2\tau}\right)}{\lambda_1\lambda_2(\lambda_1+\lambda_2)}\Bigg/\left(E\gamma+\tau+\right.$$

$$\left.-E\beta_1\left(1-e^{-\lambda_1\tau}\right)+E\beta_2\left(1-e^{-\lambda_2\tau}\right)-E(\beta_1\wedge\beta_2)\left(1-e^{-\lambda_1 t}\right)\left(1-e^{-\lambda_2 t}\right)\right).\tag{3.61}$$

Average specific income and average specific expenses are given by:

$$S=\left((c_1+c_4)\frac{\lambda_2^2\left(1-e^{-\lambda_1\tau}\right)+\lambda_1^2\left(1-e^{-\lambda_2\tau}\right)+\lambda_1\lambda_2\left(1-e^{-\lambda_1\tau}\right)\left(1-e^{-\lambda_2\tau}\right)}{\lambda_1\lambda_2(\lambda_1+\lambda_2)}-\right.$$

$$-c_3E\gamma-c_4\tau-c_2\left(E\beta_1\left(1-e^{-\lambda_1\tau}\right)+E\beta_2\left(1-e^{-\lambda_2\tau}\right)+\right.$$

$$\left.\left.+E(\beta_1\wedge\beta_2)\left(e^{-\lambda_1\tau}-1\right)\left(1-e^{-\lambda_2\tau}\right)\right)\right)\Big/\left(E\gamma+\tau+\right.$$

$$\left.+E\beta_1\left(1-e^{-\lambda_1\tau}\right)+E\beta_2\left(1-e^{-\lambda_2\tau}\right)+E(\beta_1\wedge\beta_2)\left(e^{-\lambda_1 t}-1\right)\left(1-e^{-\lambda_2 t}\right)\right),\tag{3.62}$$

$$C = \Big(c_3 E\gamma + c_4\tau + c_2\big(E\beta_1\left(1-e^{-\lambda_1\tau}\right) + E\beta_2\left(1-e^{-\lambda_2\tau}\right) -$$

$$-E(\beta_1 \wedge \beta_2)\left(1-e^{-\lambda_1\tau}\right)\left(1-e^{-\lambda_2\tau}\right)\big)\Big) \Bigg/ \left(\frac{\lambda_2^2\left(1-e^{-\lambda_1\tau}\right) + \lambda_1^2\left(1-e^{-\lambda_2\tau}\right)}{\lambda_1\lambda_2(\lambda_1+\lambda_2)} + \right.$$

$$\left. + \frac{\lambda_1\lambda_2\left(1-e^{-\lambda_1\tau}\right)\left(1-e^{-\lambda_2\tau}\right)}{\lambda_1\lambda_2(\lambda_1+\lambda_2)} \right) - c_4. \quad (3.63)$$

Example. Initial data and calculation results are included in Table 3.4. Average RT are $E\beta_1 = 0.100$ h, $E\beta_2 = 0.060$ h control duration is $E\gamma = 0.125$ h, $c_1 = 5$ c.u., $c_2 = 4$ c.u., $c_3 = 3$ c.u., $c_4 = 2$ c.u.

TABLE 3.4 Values of $K_a(\tau)$, $S(\tau)$, $C(\tau)$ Under $\tau = 5$ h

Initial data		Calculation results		
$E\alpha_1$, h	$E\alpha_2$, h	$K_a(\tau)$	$S(\tau)$, c.u./h	$C(\tau)$, c.u./h
90	70	0.9723	4.756	0.109
90	50	0.9716	4.75	0.111
90	10	0.9625	4.682	0.135

3.5 APPROXIMATION OF STATIONARY CHARACTERISTICS OF TWO-COMPONENT SERIAL SYSTEMS WITH COMPONENTS DEACTIVATION WHILE CONTROL EXECUTION

3.5.1 System Description

The system S of two serial components κ_1, κ_2 and of control unit is considered. At the initial moment, the components begin to operate, control unit is on. Components TF are RV α_1 and α_2 with DF $F_1(t) = P\{\alpha_1 \le t\}$, $F_2(t) = P\{\alpha_2 \le t\}$ and DD $f_1(t)$, $f_2(t)$, respectively. Control is carried out in a random time δ with DF $R(t) = P\{\delta \le t\}$ and DD $r(t)$. The components operability control takes place simultaneously. Failures are detected after control only. While control execution, both components are off. Control duration is RV γ with DF $V(t) = P\{\gamma \le t\}$ and DD $v(t)$. After the component κ_1 failure detection, its restoration begins, κ_2 and control unit are off. Component κ_1 RT is RV β_1 with DF $G_1(t) = P\{\beta_1 \le t\}$ and DD $g_1(t)$. After the component κ_2 failure detection, its restoration begins, κ_1 and control unit are off. Component κ_2 RT is RV β_2 with DF $G_2(t)$

$= P\{\beta_2 \le t\}$ and DD $g_2(t)$. In case of both components restoration, the system starts operating after the restoration of the latter one. All the properties of components are completely renewed after their restoration.

RV α_1, α_2, β_1, β_2, δ, γ are assumed to be independent and to have finite expectations.

Furthermore, the above system will be referred to as initial.

3.5.2 Semi-Markov Model Building of the Initial System

To describe the system S operation, let us introduce the following set E of system semi-Markov states:

$$E = \{3111, 3\hat{1}\hat{1}0x_1x_2, 3111x_1x_2, 1011x\ z, 2101x\ z, 3\hat{0}\hat{1}0x, 3\hat{1}\hat{0}0x,$$
$$32\hat{1}2x, 3\hat{1}22x, 1111x, 2111x, 1001z, 2001z, 3\hat{0}\hat{0}0, 3222, 1\hat{1}22x, 22\hat{1}2x\}.$$

Consider the sense of state codes:

3111 – components κ_1, κ_2 have been restored and begin to operate, control is on;

$3\hat{1}\hat{1}0x_1x_2$ – control has begun, components κ_1, κ_2 are operable and are off, times $x_1 > 0$ and $x_2 > 0$ are left till components κ_1 and κ_2 failures, respectively (with no regard to control execution time);

$3111x_1x_2$ – control has ended, operable components κ_1, κ_2 continue to operate; times $x_1 > 0$, $x_2 > 0$ are left till their failures;

$1011x\ z$ – component κ_1 has failed, component κ_2 is in up-state, time $x_2 > 0$ is left till its failure, time $z > 0$ is left till the control beginning;

$2101x\ z$ – component κ_2 has failed, component κ_1 is in up-state, time $x_1 > 0$ is left till its failure, time $z > 0$ is left till the control beginning;

$3\hat{0}\hat{1}0x$ – control has begun, components κ_1, κ_2 are off; component κ_1 is in down-state, while component κ_2 is in up-state, time $x_2 > 0$ is left till its failure;

$3\hat{1}\hat{0}0x$ – control has begun, components κ_1, κ_2 are off; component κ_1 is in up-state, time $x_1 > 0$ is left till its failure, component κ_2 is in down-state;

$32\hat{1}2x$ – control has been suspended because of component κ_1 failure detection, its restoration has begun, component κ_2 is in up-state and is deactivated, time $x_2 > 0$ is left till its failure;

$3\hat{1}22x$ – control has been suspended because of component κ_2 failure detection, its restoration has begun, component κ_1 is in up-state and is deactivated, time $x_1 > 0$ is left till its failure;

$1111x$ – component κ_1 has been restored and begins to operate, κ_2 continues to operate, time $x_2 > 0$ is left till its failure, control unit is on;

$2111x$ – component κ_2 has been restored and begins to operate, κ_1 continues to operate, time $x_2 > 0$ is left till its failure, control unit is on;

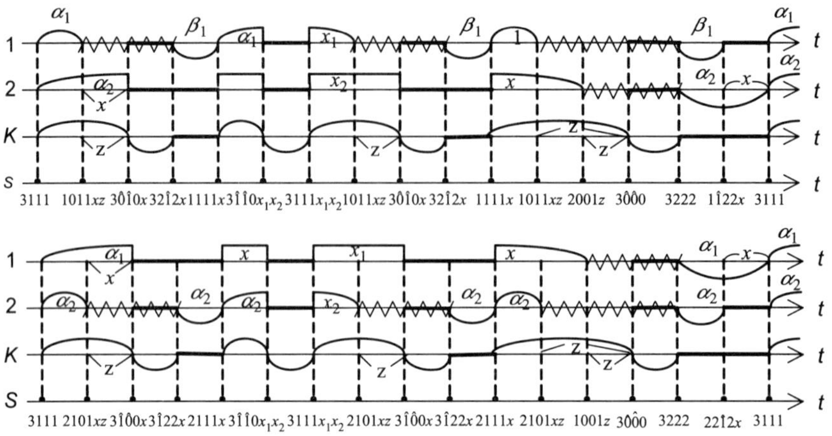

FIGURE 3.13 Time diagram of initial system operation

$1001z$ – component κ_1 has failed, component κ_2 is in down-state, time $z > 0$ is left till control execution;

$2001z$ – component κ_2 has failed, component κ_1 is in down-state, time $z > 0$ is left till control execution;

$3\hat{0}\hat{0}0$ – control has begun, components κ_1, κ_2 are in down-state and are off;

3222 – control has been carried on and suspended, failures of κ_1, κ_2 are detected, components restoration begins;

$1\hat{1}22x$ – component κ_1 has been restored and is on, time $x_2 > 0$ is left till the component κ_2 restoration, control is suspended;

$22\hat{1}2x$ – component κ_2 has been restored and is on, time $x_1 > 0$ is left till the component κ_1 restoration, control is suspended.

Time diagram of the initial system operation is given in Figure 3.13.

3.5.3 Approximation of the Initial Stationary Characteristics

To define approximate values of stationary characteristics of the initial system S, let us apply the method introduced in [14] and reviewed in Section 1.3.

Suppose, components RT and control duration is considerably less than their TF. Then the supporting system $S^{(0)}$ is the one with immediate restoration and control. Its operation is described in Section 3.1. Time diagram of the supporting system $S^{(0)}$ is in Figure 3.14, immediate states of $S^{(0)}$ are indicated in parenthesis. Transition probabilities of EMC $\{ \xi_n^{(0)}; n \geq 0 \}$ of the supporting system $S^{(0)}$ can be found by formulas (3.4), transitions from the states in parenthesis occur with the unity probability. The EMC $\{ \xi_n^{(0)}; n \geq 0 \}$ stationary distribution for the supporting system $S^{(0)}$ is obtained in Section 3.1 and is defined by (3.8).

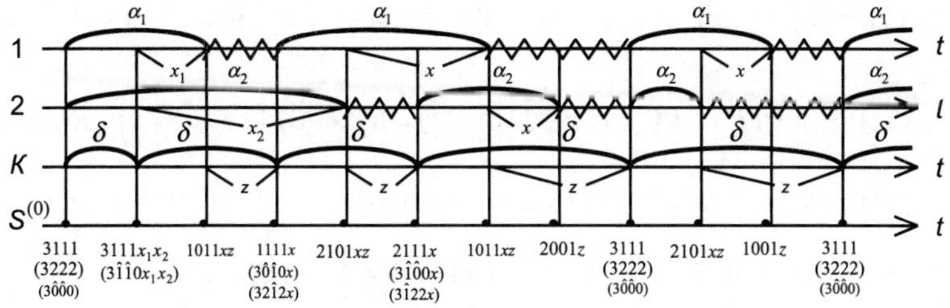

FIGURE 3.14 Time diagram of supporting system operation

The class $E^{(0)}$ of ergodic states of the supporting system EMC $\{\xi_n^{(0)}; n \geq 0\}$ is as follows:

$$E^{(0)} = \{3111, 3\hat{1}\hat{1}0x_1x_2, 3111x_1x_2, 1011x\,z, 2101x\,z, 3\hat{0}\hat{1}0x, 3\hat{1}\hat{0}0x, 32\hat{1}2x, 3\hat{1}22x, 1111x, 2111x, 1001z, 2001z, 3\hat{0}\hat{0}0, 3222\}.$$

States $1\hat{1}22x$, $22\hat{1}2x$ are transient for the supporting system EMC.

Let us define the initial system EMC $\{\xi_n; n \geq 0\}$ transition probabilities:

$$
\begin{aligned}
&p_{3111}^{1011xz} = \int_0^\infty f_2(t+x)r(t+z)f_1(t)dt,\ x>0, z>0;\\
&p_{3111}^{2101xz} = \int_0^\infty f_1(t+x)r(t+z)f_2(t)dt, x>0, z>0;\\
&p_{3111}^{3\hat{1}\hat{1}0x_1x_2} = \int_0^\infty f_1(t+x_1)f_2(t+x_2)r(t)dt, x_1>0, x_2>0;\\
&p_{3111x_1x_2}^{1011x_2-x_1z} = r(x_1+z), x_1<x_2;\\
&p_{3111x_1x_2}^{2101x_1-x_2z} = r(x_2+z), x_2<x_1;\\
&p_{3111x_1x_2}^{3\hat{1}\hat{1}0x_1-tx_2-t} = r(t), x_1<x_2, 0<t<x_1;\\
&p_{3111x_1x_2}^{3\hat{1}\hat{1}0x_1-tx_2-t} = r(t), x_2<x_1, 0<t<x_2;\\
&p_{1111x}^{1011yz} = r(x-y+z)f_1(x-y), 0<y<x, z>0;\\
&p_{1111x}^{2101yz} = r(x+z)f_1(x+y), y>0, z>0;\\
&p_{1111x}^{3\hat{1}\hat{1}0y_1y_2} = r(x-y_2)f_1(x-y_2+y_1), y_1>0,\ 0<y_2<x;\\
&p_{2111x}^{1011yz} = r(x+z)f_2(x+y), y>0, z>0;\\
&p_{2111x}^{2101yz} = r(x-y+z)f_2(x-y), 0<y<x, z>0;\\
&p_{2111x}^{3\hat{1}\hat{1}0y_1y_2} = r(x-y_1)f_2(x-y_1+y_2), y_2>0, 0<y_1<x.
\end{aligned}
\tag{3.64}
$$

Let us split the phase state space E of the initial system S into the following subsets:

$E_+ = \{3111, 3111x_1x_2, 1111x, 2111x\}$ – system up-state;

$E_- = \{3\hat{1}\hat{1}0x_1x_2, 1011x\,z, 2101x\,z, 3\hat{0}\hat{1}0x, 3\hat{1}\hat{0}0x, 32\hat{1}2x, 3\hat{1}22x, 1001z, 2001z, 3\hat{0}\hat{0}0, 3222, 1\hat{1}22x, 22\hat{1}2x\}$ – system down-state.

Average sojourn times in states of the initial system are as follows:

$$m(3111)=\int_0^{\infty}\bar{F}_1(t)\bar{F}_2(t)\bar{R}(t)\,dt, m(3\hat{1}\hat{1}0x_1x_2)=E\gamma, m(3111x_1x_2)=\int_0^{x_1\wedge x_2}\bar{R}(t)\,dt,$$

$$m(1011xz)=x\wedge z, m(2101xz)=x\wedge z, m(3\hat{0}\hat{1}0x)=E\gamma, m(3\hat{1}\hat{0}0x)=E\gamma,$$

$$m(32\hat{1}2x)=E\beta_1, m(3\hat{1}22x)=E\beta_2, m(1111x)=\int_0^{x}\bar{F}_1(t)\bar{R}(t)\,dt, \tag{3.65}$$

$$m(2111x)=\int_0^{x}\bar{F}_2(t)\bar{R}(t)\,dt, m(1001z)=z, m(2001z)=z, m(3\hat{0}\hat{0}\hat{0})=E\gamma,$$

$$m(3222)=\int_0^{\infty}\bar{G}_1(t)\bar{G}_2(t)\,dt, m(1\hat{1}22x)=x, m(22\hat{1}2x)=x.$$

Average stationary operation TF T_+ and average restoration time T_- is obtained by means of Chapter 1 (1.25 and 1.26):

$$T_+\approx\frac{\int_{E_+}m(e)\rho(de)}{\int_{E_+}P(e,E_-)\rho(de)}, T_-\approx\frac{\int_{E_-}m(e)\rho(de)}{\int_{E_+}P(e,E_-)\rho(de)}, \tag{3.66}$$

where $\rho(de)$ is the supporting EMC $\{\xi_n^{(0)};n\geq 0\}$ stationary distribution; $m(e)$ are average sojourn times in the state $e\in E$ of the initial system; $P(e,E_-)$ are the EMC $\{\xi_n;\ n\geq 0\}$ transition probabilities of the initial system. With regard to formulas (3.8), (3.64), and (3.65), we can get the expressions included in (3.66).

$$\int_{E+}m(e)\rho(de)=m(3111)\rho(3111)+\int_0^{\infty}m(1111x)\rho(1111x)\,dx+$$
$$+\int_0^{\infty}m(2111x)\rho(2111x)\,dx+\int_0^{\infty}dx_1\int_0^{\infty}m(3111x_1x_2)\rho(3111x_1x_2)\,dx_2. \tag{3.67}$$

Taking into account the transformations made in Subsection 3.1, we have the expression (3.67) in the form:

$$\int_{E_+}m(e)\rho(de)=\rho_0\left(E(\alpha_1\wedge\alpha_2)+\int_0^{\infty}\bar{F}_2(t)\bar{\Phi}_1(t)\,dt+\int_0^{\infty}\bar{F}_1(t)\bar{\Phi}_2(t)\,dt\right). \tag{3.68}$$

Here,

$$\bar{\Phi}_i(t)=\int_0^{\infty}\bar{F}_i(t+y)h_{\bar{i}}(y)\,dy+\int_0^{\infty}\bar{\Gamma}_i(t,y)\,dy\int_0^{\infty}f_i(y+z)h_i(z)\,dz+$$
$$+\int_0^{\infty}\bar{\Pi}_i(t,y)\,dy\int_0^{\infty}f_i(y+z)h_{\bar{i}}(z)\,dz+\int_0^{\infty}\bar{\Pi}_i(t,y)\,dy\int_0^{\infty}\gamma_i(y,z)\,dz\int_0^{\infty}f_{\bar{i}}(z+s)h_i(s)\,ds,$$

where

$$\bar{\Gamma}_i(t,y)=\int_t^{\infty}\gamma_i(x,y)\,dx,\bar{\Pi}_i(t,y)=\int_t^{\infty}\pi_i(x,y)\,dx,$$

$h_r(t)=\sum_{n=1}^{\infty} r^{*(n)}(t)$ is the density of renewal function $H_r(t)$ of the renewal process generated by RV δ;

$v_r(z,x)=r(z+x)+\int_0^z r(z+x-s)h_r(s)\,ds$ is the distribution density of the direct residual time for the renewal process generated by RV δ;

$h_i(t)=\sum_{n=1}^{\infty}\tilde{\gamma}_i^{*(n)}(t)$, $i=1,2$ are densities of renewal functions of renewal processes generated by RV with densities $\tilde{\gamma}_i(t)=\int_0^t f_i(t)v_r(y,t-y)\,dy$;

$$\gamma_i(x,t)=\int_0^{\infty} f_i(x+z+t)v_r(t,z)\,dz+\int_0^{\infty} h_{\bar{i}}(y)\,dy\int_0^{\infty} f_i(x+z+y+t)v_r(t,z)\,dz,\quad i=1,2;$$

$$\pi_i(x,y)=\sum_{n=1}^{\infty}k_i^{(n)}(x,y),\quad i=1,2.$$

where

$$k_i^{(1)}(x,y)=k_i(x,y)=\int_0^{\infty}\gamma_i(x,t)\gamma_{\bar{i}}(t,y)\,dt,k_i^{(n)}(x,y)=\int_0^{\infty}k_i(x,t)k_i^{(n-1)}(t,y)\,dt.$$

Next,

$$\begin{aligned}
\int_{E_-} m(e)\rho(de) &= \int_0^{\infty} dx_1\int_0^{\infty} m(3\hat{1}\hat{1}0x_1x_2)\rho(3\hat{1}\hat{1}0x_1x_2)\,dx_2+\\
&+\int_0^{\infty} dx\int_0^{\infty} m(1011x)\rho(1011xz)\,dz+\int_0^{\infty} dx\int_0^{\infty} m(2101xz)\rho(2101xz)\,dz+\\
&+\int_0^{\infty} m(3\hat{0}\hat{1}0x)\rho(3\hat{0}\hat{1}0x)\,dx+\int_0^{\infty} m(3\hat{1}\hat{0}0x)\rho(3\hat{1}\hat{0}0x)\,dx+\\
&+\int_0^{\infty} m(32\hat{1}2x)\rho(32\hat{1}2x)\,dx+\int_0^{\infty} m(3\hat{1}22x)\rho(3\hat{1}22x)\,dx+\\
&+\int_0^{\infty} m(1001z)\rho(1001z)\,dz+\int_0^{\infty} m(2001z)\rho(2001z)\,dz+\\
&+m(3\hat{0}\hat{0}\hat{0})\rho(3\hat{0}\hat{0}\hat{0})+m(3222)\rho(3222).
\end{aligned}\tag{3.69}$$

With regard to transformations made in Subsection 3.1, the expression (3.69) looks like:

$$\int_{E_-} m(e)\rho(de)=\rho_0\left((E\delta+E\gamma)\left(\int_0^\infty \bar{F}_1(y)\bar{F}_2(y)\,d\hat{H}_r(y)+\int_0^\infty \bar{\Phi}_1(y)\bar{F}_2(y)\,d\hat{H}_r(y)+\right.\right.$$

$$\left.+\int_0^\infty \bar{\Phi}_2(y)\bar{F}_1(y)\,d\hat{H}_r(y)\right)+E(\beta_1\wedge\beta_2)+E\beta_1\bar{\Phi}_2(0)+E\beta_2\bar{\Phi}_1(0)-$$

$$\left.-E(\alpha_1\wedge\alpha_2)-\int_0^\infty \bar{\Phi}_1(y)\bar{F}_2(y)\,dy-\int_0^\infty \bar{\Phi}_2(y)\bar{F}_1(y)\,dy\right). \tag{3.70}$$

Here,

$$\bar{\Phi}_i(0)=\int_0^\infty \bar{F}_i(y)h_{\bar{i}}(y)\,dy+\int_0^\infty \bar{\Gamma}_i(0,y)\,dy\int_0^\infty f_{\bar{i}}(y+z)h_i(z)\,dz+$$

$$+\int_0^\infty \bar{\Pi}_i(0,y)\,dy\int_0^\infty f_i(y+z)h_{\bar{i}}(z)\,dz+\int_0^\infty \bar{\Pi}_i(0,y)\,dy\int_0^\infty \gamma_i(y,z)\,dz\int_0^\infty f_{\bar{i}}(z+s)h_i(s)\,ds,$$

where

$$\bar{\Gamma}_i(0,y)=\int_0^\infty \gamma_i(x,y)\,dx,\ \bar{\Pi}_i(0,y)=\int_0^\infty \pi_i(x,y)\,dx,\ i=1,2.$$

Next,

$$\int_{E_+} P(e,E_-)\rho(de)=\rho_0\int_0^\infty r(t)\,dt\int_0^\infty f_1(x_1+t)\,dx_1\int_0^\infty f_2(x_2+t)\,dx_2+$$

$$+\rho_0\int_0^\infty f_1(t)\,dt\int_0^\infty f_2(x+t)\,dx\int_0^\infty r(z+t)\,dz+\rho_0\int_0^\infty f_2(t)\,dt\int_0^\infty f_1(x+t)\,dx\int_0^\infty r(z+t)\,dz+$$

$$+\int_0^\infty dx_1\int_{x_1}^\infty \rho(3111x_1x_2)\,dx_2\int_0^{x_1} r(z)\,dz+\int_0^\infty dx_1\int_0^{x_1} \rho(3111x_1x_2)\,dx_2\int_0^{x_2} r(z)\,dz+$$

$$+\int_0^\infty dx_1\int_{x_1}^\infty \rho(3111x_1x_2)\,dx_2\int_0^\infty r(x_1+z)\,dz+\int_0^\infty dx_1\int_0^{x_1} \rho(3111x_1x_2)\,dx_2\int_0^\infty r(x_2+z)\,dz+$$

$$+\int_0^\infty \rho(1111x)\,dx\int_0^x r(x-y)\,dy\int_0^\infty f_1(x-y+t)\,dt+$$

$$+\int_0^\infty \rho(1111x)\,dx\int_0^\infty f_1(x+y)\,dy\int_0^\infty r(x+z)\,dz+$$

$$+\int_0^\infty \rho(1111x)\,dx\int_0^x f_1(x-y)\,dy\int_0^\infty r(x-y+z)\,dz+$$

$$+\int_0^\infty \rho(2111x)\,dx\int_0^\infty f_2(x+y)\,dy\int_0^\infty r(x+z)\,dz+$$
$$+\int_0^\infty \rho(2111x)\,dx\int_0^x f_2(x-y)\,dy\int_0^\infty r(x-y+z)\,dz+$$
$$+\int_0^\infty \rho(2111x)\,dx\int_0^x r(x-y)\,dy\int_0^\infty f_2(x-y+t)\,dt. \tag{3.71}$$

Taking into account that $\rho(3111)=\rho_0$, $\varphi_1(x)=\rho(2111x)$, $\varphi_2(x)=\rho(1111x)$, $\varphi_3(x_1,x_2)=\rho(3111x_1x_2)$, we can rewrite formula (3.71) with the help of simplifications from Subsection 3.1:

$$\int_{E_+} P(e,E_-)\rho(de)=\rho_0\left(\int_0^\infty \bar{F}_1(y)\bar{F}_2(y)\,d\hat{H}_r(y)+\right.$$
$$\left.+\int_0^\infty \bar{\Phi}_1(y)\bar{F}_2(y)\,d\hat{H}_r(y)+\int_0^\infty \bar{\Phi}_2(y)\bar{F}_1(y)\,d\hat{H}_r(y)\right) \tag{3.72}$$

In such a way, with regard to (3.68) and (3.72), approximate formula for average stationary operating TF T_+ is:

$$T_+\approx\frac{E(\alpha_1\wedge\alpha_2)+\int_0^\infty \bar{F}_2(t)\bar{\Phi}_1(t)\,dt+\int_0^\infty \bar{F}_1(t)\bar{\Phi}_2(t)\,dt}{\int_0^\infty \bar{F}_1(y)\bar{F}_2(y)\,d\hat{H}_r(y)+\int_0^\infty \bar{\Phi}_2(y)\bar{F}_1(y)\,d\hat{H}_r(y)+\int_0^\infty \bar{\Phi}_1(y)\bar{F}_2(y)\,d\hat{H}_r(y)}. \tag{3.73}$$

With regard to (3.70) and (3.72), approximate formula for average stationary restoration time T_- is:

$$T_-\approx\left((E\delta+E\gamma)\left(\int_0^\infty \bar{F}_1(y)\bar{F}_2(y)\,d\hat{H}_r(y)+\int_0^\infty \bar{\Phi}_1(y)\bar{F}_2(y)\,d\hat{H}_r(y)+\right.\right.$$
$$\left.+\int_0^\infty \bar{\Phi}_2(y)\bar{F}_1(y)\,d\hat{H}_r(y)\right)+E(\beta_1\wedge\beta_2)+E_1\bar{\Phi}_2(0)+E\beta_2\bar{\Phi}_1(0)-$$
$$\left.-E(\alpha_1\wedge\alpha_2)-\int_0^\infty \bar{\Phi}_1(y)\bar{F}_2(y)\,dy-\int_0^\infty \bar{\Phi}_2(y)\bar{F}_1(y)\,dy\right)\Big/$$
$$\left(\int_0^\infty \bar{F}_1(y)\bar{F}_2(y)\,d\hat{H}_r(y)+\int_0^\infty \bar{\Phi}_1(y)\bar{F}_2(y)\,d\hat{H}_r(y)+\int_0^\infty \bar{\Phi}_2(y)\bar{F}_1(y)\,d\hat{H}_r(y)\right). \tag{3.74}$$

For the stationary availability factor, with regard to (3.73) and (3.74), the following approximate ratio takes place:

$$K_a \approx \left(E(\alpha_1 \wedge \alpha_2) + \int_0^\infty \bar{F}_2(t)\bar{\Phi}_1(t)\,dt + \int_0^\infty \bar{F}_1(t)\bar{\Phi}_2(t)\,dt \right) \Big/ \Big((E\delta + E\gamma) \times$$
$$\times \left(\int_0^\infty \bar{F}_1(y)\bar{F}_2(y)\,d\hat{H}_r(y) + \int_0^\infty \bar{\Phi}_1(y)\bar{F}_2(y)\,d\hat{H}_r(y) + \right.$$
$$\left. \int_0^\infty \bar{\Phi}_2(y)\bar{F}_1(y)\,d\hat{H}_r(y) \right) + E(\beta_1 \wedge \beta_2) + E\beta_1\bar{\Phi}_2(0) + E\beta_2\bar{\Phi}_1(0) \Big). \quad (3.75)$$

Let us obtain initial system stationary efficiency characteristics: average specific income S per calendar time unit and average specific expenses C per time unit of system up-state. To approximate them we apply as given in Chapter 1 (1.27).

Let c_1 be the income per time unit of the initial system up-state, c_2 are expenses per time unit of control, c_3 are expenses per time unit of components restoration; c_4 are expenses per time unit of latent failure. For the initial system, the functions $f_s(e), f_c(e)$ are:

$$f_s(e) = \begin{cases} c_1, & e \in \{3111,\ \ 3111x_1x_2,\ \ 2111x,\ \ 1111x\}, \\ -c_2, & e \in \{32\hat{1}2x,\ \ 3\hat{1}22x,\ \ 3222,\ \ 1\hat{1}22x,\ \ 22\hat{1}2x\}, \\ -c_3, & e \in \{3\hat{1}\hat{1}0x_1x_2,\ \ 3\hat{0}\hat{1}0x,\ \ 3\hat{1}\hat{0}0x,\ \ 3\hat{0}\hat{0}0\}, \\ -c_4, & e \in \{1011x\ \ z,\ \ 2101x\ \ z,\ \ 1001z,\ \ 2001z\}, \end{cases}$$
$$f_c(e) = \begin{cases} 0, & e \in \{3111,\ \ 3111x_1x_2,\ \ 2111x,\ \ 1111x\}, \\ c_2, & e \in \{32\hat{1}2x,\ \ 3222,\ \ 3\hat{1}22x\}, \\ c_3, & e \in \{3\hat{1}\hat{1}0x_1x_2,\ \ 3\hat{0}\hat{1}0x,\ \ 3\hat{1}00x,\ \ 3\hat{0}\hat{0}0\}, \\ c_4, & e \in \{1011x\ \ z,\ \ 2101x\ \ z,\ \ 1001z,\ \ 2001z\}. \end{cases} \quad (3.76)$$

Applying formulas (3.5), (3.65), and (3.76), let us obtain the expressions from Chapter 1 (1.27):

$$\int_E m(e)f_s(e)\rho(de) = c_1 \left(\rho(3111)m(3111) + \int_0^\infty m(2111x)\rho(2111x)\,dx + \right.$$
$$\left. + \int_0^\infty m(1111x)\rho(1111x)\,dx + \int_0^\infty dx_1 \int_0^\infty m(3111x_1x_2)\rho(3111x_1x_2)\,dx_2 \right) -$$
$$-c_3 \left(\int_0^\infty m(3\hat{0}\hat{1}0x)\rho(3\hat{0}\hat{1}0x)\,dx + \int_0^\infty \int_0^\infty m(3\hat{1}\hat{1}0x_1x_2)\rho(3\hat{1}\hat{1}0x_1x_2)\,dx_1\,dx_2 + \right.$$
$$\left. + m(3\hat{0}\hat{0}0)\rho(3\hat{0}\hat{0}0) + \int_0^\infty m(3\hat{1}\hat{0}0x)\rho(3\hat{1}\hat{0}0x)\,dx \right) -$$

$$
\begin{aligned}
&-c_2\left(\int_0^\infty m(32\hat{1}2x)\rho(32\hat{1}2x)\,dx + m(3222)\rho(3222) + \int_0^\infty m(3\hat{1}22x)\rho(3\hat{1}22x)\,dx\right) - \\
&-c_4\left(\int_0^\infty dx\int_0^\infty m(1011xz)\rho(1011xz)\,dz + \int_0^\infty m(2001z)\rho(2001z)\,dz + \right.\\
&\left. + \int_0^\infty dx\int_0^\infty m(2101xz)\rho(2101xz)\,dz + \int_0^\infty m(1001z)\rho(1001z)\,dz\right) = \\
&= \rho_0\left((c_1+c_4)\left(E(\alpha_1\wedge\alpha_2) + \int_0^\infty \overline{F}_2(t)\overline{\Phi}_1(t)\,dt + \int_0^\infty \overline{F}_1(t)\overline{\Phi}_2(t)\,dt\right) - \right.\\
&-(c_3E\gamma + c_4E\delta)\left(\int_0^\infty \overline{F}_1(y)\overline{F}_2(y)\,d\hat{H}_r(y) + \int_0^\infty \overline{\Phi}_1(y)\overline{F}_2(y)\,d\hat{H}_r(y) + \right.\\
&\left.\left. + \int_0^\infty \overline{\Phi}_2(y)\overline{F}_1(y)\,d\hat{H}_r(y)\right) - c_2\left(E(\beta_1\wedge\beta_2) + E\beta_1\overline{\Phi}_2(0) + E\beta_2\overline{\Phi}_1(0)\right)\right).
\end{aligned}
\tag{3.77}
$$

Next,

$$
\begin{aligned}
&\int_E m(e)f_c(e)\rho(de) = \\
&= c_3\left(\int_0^\infty\int_0^\infty m(3\hat{1}\hat{1}0x_1x_2)\rho(3\hat{1}\hat{1}0x_1x_2)\,dx_1\,dx_2 + \int_0^\infty m(3\hat{0}\hat{1}0x)\rho(3\hat{0}\hat{1}0x)\,dx + \right.\\
&\left. + \int_0^\infty m(3\hat{0}\hat{1}0x)\rho(3\hat{0}\hat{1}0x)\,dx + \int_0^\infty m(3\hat{1}\hat{0}0x)\rho(3\hat{1}\hat{0}0x)\,dx + m(3\hat{0}\hat{0}0)\rho(3\hat{0}\hat{0}0)\right) + \\
&+ c_2\left(\int_0^\infty m(32\hat{1}2x)\rho(32\hat{1}2x)\,dx + \int_0^\infty m(3\hat{1}22x)\rho(3\hat{1}22x)\,dx + m(3222)\rho(3222)\right) + \\
&+ c_4\left(\int_0^\infty dx\int_0^\infty m(1011xz)\rho(1011xz)\,dz + \int_0^\infty m(2001z)\rho(2001z)\,dz + \right.\\
&\left. + \int_0^\infty dx\int_0^\infty m(2101xz)\rho(2101xz)\,dz + \int_0^\infty m(1001z)\rho(1001z)\,dz\right) = \\
&= \rho_0\left((c_3E\gamma + c_4E\delta)\left(\int_0^\infty \overline{F}_1(y)\overline{F}_2(y)\,d\hat{H}_r(y) + \int_0^\infty \overline{\Phi}_1(y)\overline{F}_2(y)\,d\hat{H}_r(y) + \right.\right.\\
&\left. + \int_0^\infty \overline{\Phi}_2(y)\overline{F}_1(y)\,d\hat{H}_r(y)\right) + c_2\left(E(\beta_1\wedge\beta_2) + E\beta_1\overline{\Phi}_2(0) + E\beta_2\overline{\Phi}_1(0)\right) - \\
&\left. - c_4\left(E(\alpha_1\wedge\alpha_2) + \int_0^\infty \overline{F}_2(t)\overline{\Phi}_1(t)\,dt + \int_0^\infty \overline{F}_1(t)\overline{\Phi}_2(t)\,dt\right)\right).
\end{aligned}
\tag{3.78}
$$

With regard to (3.67), (3.69), (3.76), and (3.77), average specific income and average specific expenses are as follows:

$$S \approx \left((c_1 + c_4)\left(E(\alpha_1 \wedge \alpha_2) + \int_0^\infty \bar{F}_2(t)\bar{\Phi}_1(t)dt + \int_0^\infty \bar{F}_1(t)\bar{\Phi}_2(t)dt \right) - \right.$$
$$- (c_3 E\gamma + c_4 E\delta)\left(\int_0^\infty \bar{F}_1(y)\bar{F}_2(y)d\hat{H}_r(y) + \int_0^\infty \bar{\Phi}_1(y)\bar{F}_2(y)d\hat{H}_r(y) + \right.$$
$$\left. + \int_0^\infty \bar{\Phi}_2(y)\bar{F}_1(y)d\hat{H}_r(y) \right) - c_2\left(E(\beta_1 \wedge \beta_2) + E\beta_1\bar{\Phi}_2(0) + \right.$$
$$\left.\left. + E\beta_2\bar{\Phi}_1(0)\right)\right) \Big/ \left((E\gamma + E\delta)\left(\int_0^\infty \bar{F}_1(y)\bar{F}_2(y)d\hat{H}_r(y) + \int_0^\infty \bar{\Phi}_1(y)\bar{F}_2(y)d\hat{H}_r(y) + \right.\right.$$
$$\left.\left. + \int_0^\infty \bar{\Phi}_2(y)\bar{F}_1(y)d\hat{H}_r(y) \right) + E(\beta_1 \wedge \beta_2) + E\beta_1\bar{\Phi}_2(0) + E\beta_2\bar{\Phi}_1(0) \right), \tag{3.79}$$

$$C \approx \left((c_3 E\gamma + c_4 E\delta)\left(\int_0^\infty \bar{F}_1(y)\bar{F}_2(y)d\hat{H}_r(y) + \int_0^\infty \bar{\Phi}_1(y)\bar{F}_2(y)d\hat{H}_r(y) + \right.\right.$$
$$\left. + \int_0^\infty \bar{\Phi}_2(y)\bar{F}_1(y)d\hat{H}_r(y) \right) + c_2\left(E(\beta_1 \wedge \beta_2) + E\beta_1\bar{\Phi}_2(0) + E\beta_2\bar{\Phi}_1(0) \right) -$$
$$\left.\left. - c_4\left(E(\alpha_1 \wedge \alpha_2) + \int_0^\infty \bar{F}_2(t)\bar{\Phi}_1(t)dt + \int_0^\infty \bar{F}_1(t)\bar{\Phi}_2(t)dt \right)\right)\right) \Big/ \left(E(\alpha_1 \wedge \alpha_2 + \right.$$
$$\left. + \int_0^\infty \bar{F}_2(t)\bar{\Phi}_1(t)dt + \int_0^\infty \bar{F}_1(t)\bar{\Phi}_2(t)dt \right). \tag{3.80}$$

Consider the case of exponential distribution of components TF. Then $F_i(t) = 1 - e^{-\lambda_i t}$, where $i = 1, 2$, $R(t) = 1 - e^{-\mu t}$.

Let us get system stationary characteristics.

$$\int_{E+} m(e)\rho(de) = \rho_0 \frac{(\lambda_1 + \mu)(\lambda_2 + \mu)}{\lambda_1\lambda_2(\lambda_1 + \lambda_2 + 2\mu)},$$

$$\int_{E_-} m(e)\rho(de) = \rho_0 \left(\frac{(\lambda_1 + \mu)(\lambda_2 + \mu)(\lambda_1 + \lambda_2)}{\lambda_1\lambda_2\mu(\lambda_1 + \lambda_2 + 2\mu)} + E\gamma \frac{(\lambda_1 + \mu)(\lambda_2 + \mu)(\lambda_1 + \lambda_2 + \mu)}{\lambda_1\lambda_2(\lambda_1 + \lambda_2 + 2\mu)} + \right.$$
$$\left. + \frac{E\beta_1\lambda_1\mu(\lambda_1 + \mu) + E\beta_2\lambda_2\mu(\lambda_2 + \mu) + E(\beta_1 \wedge \beta_2)\lambda_1\lambda_2(\lambda_1 + \lambda_2 + 2\mu)}{\lambda_1\lambda_2(\lambda_1 + \lambda_2 + 2\mu)} \right),$$

$$\int_{E_+} P(e, E_-)\rho(de) = \rho_0 \frac{(\lambda_1 + \mu)(\lambda_2 + \mu)(\lambda_1 + \lambda_2 + \mu)}{\lambda_1\lambda_2(\lambda_1 + \lambda_2 + 2\mu)}.$$

Then average stationary operating TF T_+ and average stationary restoration time T_- are:

$$T_+ \approx \frac{E\alpha_1 E\alpha_2 E\delta}{E\alpha_1 E\delta + E\alpha_2 E\delta + E\alpha_1 E\alpha_2},$$
$$T_- \approx E\gamma + \frac{E\delta(\lambda_1+\lambda_2)}{\lambda_1+\lambda_2+\mu} + \frac{E\beta_1\lambda_1\mu}{(\lambda_2+\mu)(\lambda_1+\lambda_2+\mu)} +$$
$$+ \frac{E\beta_2\lambda_2\mu}{(\lambda_1+\mu)(\lambda_1+\lambda_2+\mu)} + \frac{E(\beta_1\wedge\beta_2)\lambda_1\lambda_2(\lambda_1+\lambda_2+2\mu)}{(\lambda_1+\mu)(\lambda_2+\mu)(\lambda_1+\lambda_2+\mu)}.$$

Stationary availability factor takes the following form:

$$K_a \approx \frac{E\alpha_1 E\alpha_2 E\delta}{E\alpha_1 E\alpha_2 + E\alpha_1 E\delta + E\alpha_2 E\delta} \Bigg/ \left(E\gamma + E\delta + \frac{E\beta_1\lambda_1\mu}{(\lambda_2+\mu)(\lambda_1+\lambda_2+\mu)} + \right.$$
$$\left. + \frac{E\beta_2\lambda_2\mu}{(\lambda_1+\mu)(\lambda_1+\lambda_2+\mu)} + \frac{E(\beta_1\wedge\beta_2)\lambda_1\lambda_2(\lambda_1+\lambda_2+2\mu)}{(\lambda_1+\mu)(\lambda_2+\mu)(\lambda_1+\lambda_2+\mu)} \right).$$

Average specific income and average specific expenses are determined by the following expressions:

$$S \approx \big((c_1+c_4)(\lambda_1+\mu)(\lambda_2+\mu) - (c_3E\gamma+c_4)(\lambda_1+\mu)(\lambda_2+\mu)(\lambda_1+\lambda_2+\mu) -$$
$$-c_2\big(E\beta_1\lambda_1\mu(\lambda_1+\mu) + E\beta_2\lambda_2\mu(\lambda_2+\mu) +$$
$$+E(\beta_1\wedge\beta_2)\lambda_1\lambda_2(\lambda_1+\lambda_2+2\mu)\big)\big) \big/ \big((E\gamma+E\delta)(\lambda_1+\mu)(\lambda_2+\mu)(\lambda_1+\lambda_2+\mu) +$$
$$+E\beta_1\lambda_1\mu(\lambda_1+\mu) + E\beta_2\lambda_2\mu(\lambda_2+\mu) + E(\beta_1\wedge\beta_2)\lambda_1\lambda_2(\lambda_1+\lambda_2+2\mu)\big),$$
$$C \approx (c_3E\gamma + c_4E\delta)(\lambda_1+\lambda_2+\mu) +$$
$$+c_2\frac{E\beta_1\lambda_1\mu(\lambda_1+\mu) + E\beta_2\lambda_2\mu(\lambda_2+\mu) + E(\beta_1\wedge\beta_2)\lambda_1\lambda_2(\lambda_1+\lambda_2+2\mu)}{(\lambda_1+\mu)(\lambda_2+\mu)(\lambda_1+\lambda_2+\mu)} - c_4.$$

Consider the case of nonrandom control periodicity $\tau > 0$. In this case, $R(t) = 1(t - \tau)$, $E\delta = \tau$. That is why approximate values of average stationary operating TF T_+, average stationary restoration time T_-, stationary availability factor, average specific income and average specific expenses are:

$$T_+ \approx \frac{1-e^{-(\lambda_1+\lambda_2)\tau}}{(\lambda_1+\lambda_2)},$$
$$T_- \approx E\gamma + \tau - \frac{1-e^{-(\lambda_1+\lambda_2)\tau}}{(\lambda_1+\lambda_2)} + E\beta_1 e^{-\lambda_2\tau}\left(1-e^{-\lambda_1\tau}\right) +$$
$$+E\beta_2 e^{-\lambda_1\tau}\left(1-e^{-\lambda_2\tau}\right) + E(\beta_1\wedge\beta_2)(1-e^{-\lambda_1\tau})(1-e^{-\lambda_2\tau}), \tag{3.81}$$
$$K_a \approx \left(\left(1-e^{-(\lambda_1+\lambda_2)\tau}\right)/(\lambda_1+\lambda_2)\right) \big/ \big(E\gamma + \tau + E\beta_1 e^{-\lambda_2\tau}(1-e^{-\lambda_1\tau}) +$$
$$+E\beta_2 e^{-\lambda_1\tau}(1-e^{-\lambda_2\tau}) + E(\beta_1\wedge\beta_2)(1-e^{-\lambda_1\tau})(1-e^{-\lambda_2\tau})\big),$$

$$S \approx \left((c_1 + c_4)\frac{1-e^{-(\lambda_1+\lambda_2)\tau}}{(\lambda_1+\lambda_2)} - (c_3 E\gamma + c_4\tau) - c_2\left(E\beta_1 e^{-\lambda_2\tau}\left(1-e^{-\lambda_1\tau}\right)+ \right.\right.$$
$$\left. +E\beta_2 e^{-\lambda_1\tau}(1-e^{-\lambda_2\tau}) + E(\beta_1 \wedge \beta_2)(1-e^{-\lambda_1\tau})(1-e^{-\lambda_2\tau})\right)\Big/\left(E\gamma + \tau + \right.$$
$$\left.\left. + E\beta_1 e^{-\lambda_2\tau}(1-e^{-\lambda_1\tau}) + E\beta_2 e^{-\lambda_1\tau}(1-e^{-\lambda_2\tau}) + E(\beta_1 \wedge \beta_2)(1-e^{-\lambda_1\tau})(1-e^{-\lambda_2\tau})\right)\right), \tag{3.82}$$

$$C \approx (c_3 E\gamma + c_4\tau)\frac{(\lambda_1+\lambda_2)}{(1-e^{-(\lambda_1+\lambda_2)\tau})} + c_2\frac{(\lambda_1+\lambda_2)}{(1-e^{-(\lambda_1+\lambda_2)\tau})}\left(E\beta_1 e^{-\lambda_2\tau}\left(1-e^{-\lambda_1\tau}\right)+\right.$$
$$\left. +E\beta_2 e^{-\lambda_1\tau}(1-e^{-\lambda_2\tau}) + E(\beta_1 \wedge \beta_2)(1-e^{-\lambda_1\tau})(1-e^{-\lambda_2\tau})\right) - c_4. \tag{3.83}$$

Example. Initial data and calculation results are given in Table 3.5, average RT are $E\beta_1 = 0.100$ h, $E\beta_2 = 0.060$ h, control duration is $E\gamma = 0.125$ h, $c_1 = 5$ c.u., $c_2 = 4$ c.u., $c_3 = 3$ c.u., $c_4 = 2$ c.u.

Using data from Tables 3.3 and 3.5, one can evaluate the error of stationary characteristics approximate calculations. Comparative results are represented in Table 3.6.

TABLE 3.5 Approximate Values of $K_a(\tau)$, $S(\tau)$, $C(\tau)$ Under $\tau = 5$ h

Initial data		Calculation results		
$E\alpha_1$, h	$E\alpha_2$, h	$K_a(\tau)$	$S(\tau)$, c.u./h	$C(\tau)$, c.u./h
90	70	0.915	4.352	0.242
90	50	0.902	4.262	0.274
90	10	0.746	3.165	0.756

TABLE 3.6 Comparison of Calculation Results

Initial data	$K_a(\tau)$, according to (3.49)	$K_a(\tau)$, according to (3.81)	Error (%)
$E\alpha_1 = 90$ h; $E\alpha_2 = 70$ h	0.91474	0.91476	0.002
$E\alpha_1 = 90$ h; $E\alpha_2 = 50$ h	0.90188	0.90191	0.003
$E\alpha_1 = 90$ h; $E\alpha_2 = 10$ h	0.74550	0.74559	0.009

Chapter 4

Optimization of Execution Periodicity of Latent Failures Control

Chapter Outline

4.1 DEFINITION OF OPTIMAL CONTROL PERIODICITY FOR ONE-COMPONENT SYSTEMS

In engineering systems, being undetected, latent failures result in losses. On the other hand, control execution itself requires expenses. That is why the problem of control execution periodicity optimization arises [9, 27].

The problems are concerned with the definition of control periodicity τ, ensuring absolute maxima of steady-state availability factor $K_a(\tau_{opt})$, of average specific income per calendar time unit $S(\tau^s_{opt})$, and absolute minimum of average specific expenses per time unit of system up-state $C(\tau^c_{opt})$.

To solve the problems for one-component systems, expressions of reliability and efficiency characteristics from Chapter 2 are taken as objective functions.

Semi-Markov Models. http://dx.doi.org/10.1016/B978-0-12-802212-2.00004-8
Copyright © 2015 Elsevier Inc. All rights reserved.

The optimum analysis of characteristics as the functions of single argument (control execution periodicity τ), enables to determine critical points, ensuring optimal values of objective functions.

4.1.1 Control Periodicity Optimization for One-Component System with Component Deactivation

For this system, the objective functions $K_a(\tau)$, $S(\tau)$, and $C(\tau)$ are given by (2.16), (2.17), and (2.18), respectively. The problems of definition of optimal values of stationary characteristics reduce to obtaining the points of unconditional extrema of these functions. By differentiating the functions with respect to τ and equating derivatives to zero, we get formulas (4.1), (4.2), and (4.3) to determine critical points of objective functions:

$$(\tau+E\gamma)\sum_{n=0}^{\infty} n f(n\tau)=\sum_{n=0}^{\infty}\bar{F}(n\tau), \tag{4.1}$$

$$\left(E\beta\,(c_3E\gamma+c_4\tau)+(\tau+E\gamma)\,(E\alpha(c_1+c_4)-c_2E\beta)\right)\sum_{n=0}^{\infty} n\ f(n\tau)=$$

$$=\left(E\alpha(c_1+c_4)-E\beta(c_2-c_4)-E\gamma(c_3-c_4)\sum_{n=0}^{\infty}\bar{F}(n\tau)\right)\sum_{n=0}^{\infty}\bar{F}(n\tau), \tag{4.2}$$

$$\left(c_4\tau+c_3E\gamma\right)\sum_{n=0}^{\infty} n f(n\tau)=c_4\sum_{n=0}^{\infty}\bar{F}(n\tau). \tag{4.3}$$

The solution of the equations enables to define: optimal control periodicity τ_{opt} corresponding the maximum value of stationary availability factor $K_a(\tau_{opt})$, optimal periodicity τ^s_{opt}, ensuring the maximum value of average specific income $S(\tau^s_{opt})$ per calendar time unit, and optimal periodicity τ^c_{opt}, giving the minimum of average specific expenses per $C(\tau^c_{opt})$ time unit of system up-state.

Example. Initial data for the optimal control periodicity definition are average component time to failure $E\alpha$, h; average restoration time $E\beta$, h; and average control duration $E\gamma$, h. The random value α has the distributions: exponential, Erlangian of the 4th order, and Veibull–Gnedenko with the shape parameter β=2. To optimize efficiency characteristics the following initial data are taken: c_1 = 5 c.u.; c_2 = 3 c.u.; c_3 = 2 c.u.; and c_4 = 4 c.u. The values of τ_{opt}, τ^s_{opt}, τ^c_{opt}, and corresponding values of $K_a(\tau_{opt})$, $S(\tau^s_{opt})$, $C(\tau^c_{opt})$ are give in Table 4.1.

The graphs of functions $K_a(\tau)$, $S(\tau)$, $C(\tau)$, in case of Erlangian distribution of the 4th order under $E\alpha=60$ h, $E\beta=0.5$ h, $h=E\gamma=0.2$ h, are presented in Figures 4.1, 4.2, and 4.3, respectively.

TABLE 4.1 Optimal Control Periodicity and Corresponding Values of $K_a(\tau_{opt})$, $S(\tau^s_{opt})$, $C(\tau^c_{opt})$ for One-Component System with Component Deactivation

Initial data				Results					
Kind of RV α distribution	$E\alpha$, h	$E\beta$, h	$E\gamma$, h	τ_{opt}, h	$K_a(\tau_{opt})$,	τ^s_{opt}, h	$S(\tau^s_{opt})$, c.u./h	τ^c_{opt}, h	$C(\tau^c_{opt})$, c.u./h
Exponential	60	0.5	0.2	4.833	0.916	4.221	4.332	3.431	0.26
Exponential	70	0.33	0.25	5.834	0.916	4.271	4.332	4.142	0.257
Erlangian of the 4th order	60	0.5	0.2	4.899	0.916	4.271	4.332	3.464	0.259
Erlangian of the 4th order	70	0.33	0.25	5.916	0.917	4.271	4.332	4.183	0.257
Veibull–Gnedenko	60	0.5	0.2	4.899	0.916	4.271	4.335	3.464	0.259
Veibull–Gnedenko	70	0.33	0.25	5.916	0.917	5.158	4.339	4.183	0.257

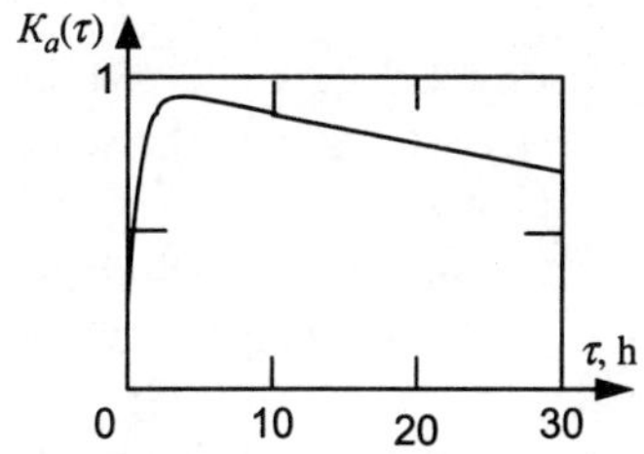

FIGURE 4.1 Graph of stationary availability factor $K_a(\tau)$ against control periodicity τ

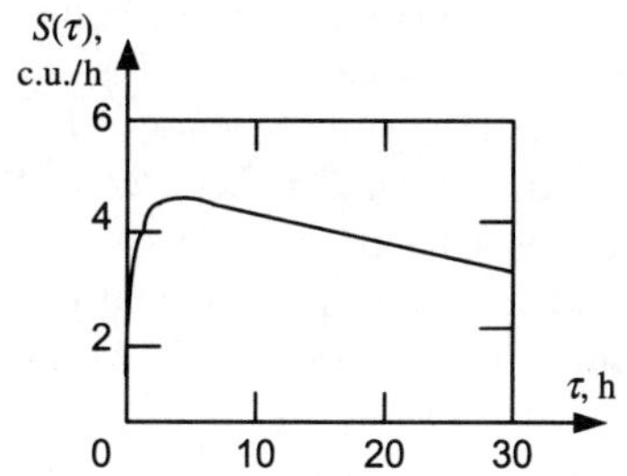

FIGURE 4.2 Graph of average specific income $S(\tau)$ against control periodicity τ

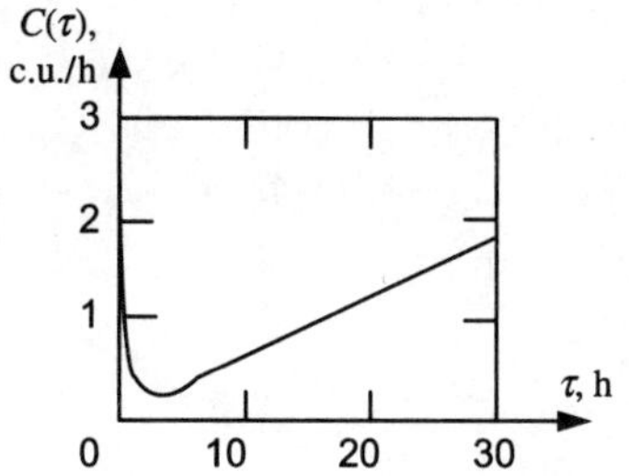

FIGURE 4.3 Graph of average specific expenses $C(\tau)$ against control periodicity τ

4.1.2 Optimal Control Periodicity for One-Component System Without Deactivation

In this case, objective functions $K_a(\tau)$, $S(\tau)$, $C(\tau)$ are determined by the formulas (2.37), (2.46), and (2.47). The equations (4.4), (4.5), and (4.6) for critical points are:

$$(\tau+h)\sum_{n=0}^{\infty} n\cdot f(n(\tau+h))=\sum_{n=0}^{\infty}\bar{F}(n(\tau+h)), \tag{4.4}$$

$$\big((\tau+h)\big(E\beta(c_2-c_4)-E\alpha(c_1+c_4)\big)-E\beta c_3 h\big)\sum_{n=0}^{\infty} n\, f(n(\tau+h))=$$

$$=\left(E\beta(c_2-c_4)-E\alpha(c_1+c_4)+hc_3\sum_{n=0}^{\infty}\bar{F}(n(\tau+h))\right)\sum_{n=0}^{\infty}\bar{F}(n(\tau+h)), \tag{4.5}$$

$$\big(c_4(\tau+h)+c_3 h\big)\sum_{n=0}^{\infty} n\cdot f(n(\tau+h))=c_4\sum_{n=0}^{\infty}\bar{F}(n(\tau+h)). \tag{4.6}$$

The results of τ^s_{opt}, τ^c_{opt} definition and corresponding values of $S(\tau^s_{opt})$, $C(\tau^c_{opt})$ for the system without component deactivation are given in Table 4.2.

To compare calculation results, initial data are taken the same as in for the system with deactivation.

The graphs of functions $S(\tau)$, $C(\tau)$, in case of Erlangian distribution of the 4th order under $E\alpha=60$ h, $E\beta=0.5$ h, $h=E\gamma=0.2$ h are represented in Figures 4.4 and 4.5.

One should note, for the present system, optimization of control periodicity can be carried out according to the objective functions $S(\tau)$ and $C(\tau)$

TABLE 4.2 Optimal Control Periodicity and Values of $S(\tau^s_{opt})$, $C(\tau^c_{opt})$

Initial data				Results			
Kind of RV α distribution	$E\alpha$, h	$E\beta$, h	h, h	τ^s_{opt}, h	$S(\tau^s_{opt})$, c.u./h	τ^c_{opt}, h	$C(\tau^c_{opt})$, c.u./h
Exponential	60	0.5	0.20	3.949	4.355	4.355	0.253
Exponential	70	0.33	0.25	6.117	5.034	5.622	0.367
Erlangian of the 4th order	60	0.5	0.20	3.994	4.358	5.074	0.259
Erlangian of the 4th order	70	0.33	0.25	7.452	6.113	5.162	0.357
Veibull–Gnedenko	60	0.5	0.20	4.125	4.985	4.982	0.319
Veibull–Gnedenko	70	0.33	0.25	5.158	4.339	4.183	0.257

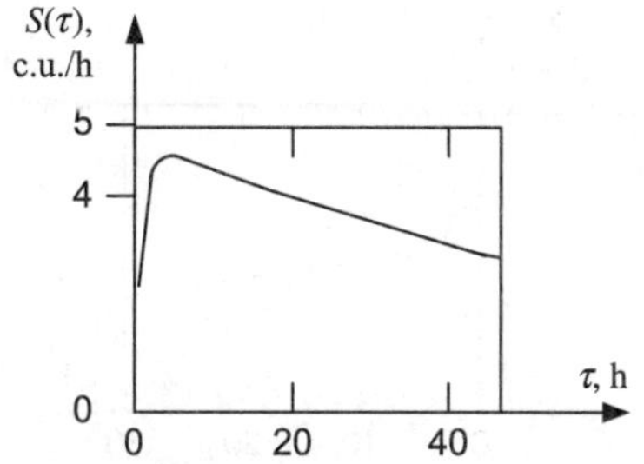

FIGURE 4.4 Graph of average specific income $S(\tau)$ against control periodicity τ

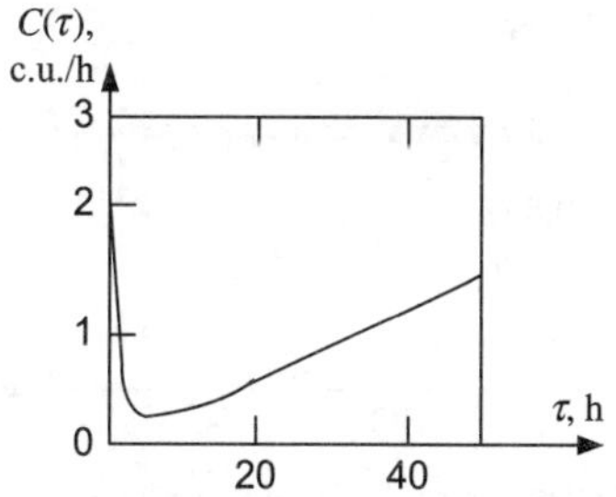

FIGURE 4.5 Graph of average specific expenses $C(\tau)$ against control periodicity τ

only. Stationary availability factor $K_a(\tau)$ takes the maximum value in case of continuous control execution, that is, under $\tau = 0$.

4.1.3 Control Periodicity Optimization for One-Component System with Regard to Component Deactivation and Control Failures

Applying formulas (2.82), (2.83), and (2.84) as objective functions allows us to determine optimal control periodicity of the system with possible control failures. The equations for critical points definition are not given here because of their length. To calculate the optimal values of periods τ_{opt}, τ^s_{opt}, τ^c_{opt} ensuring maximum of stationary availability factor $K_a(\tau_{opt})$, maximum of average specific income $S(\tau^s_{opt})$ per calendar time unit, and minimum of average specific expenses $C(\tau^c_{opt})$ per time unit of system up-state correspondingly, the following initial data are taken: component average time to failure $E\alpha$, h, average restoration time $E\beta$, h, average control duration $E\gamma$, h, $c_1 = 2$ c.u., $c_2 = 2$ c.u., $c_3 = 1$ c.u., $c_4 = 1$ c.u. RV α distributions are exponential and Erlangian of the 2nd order. Calculation results are placed in Table 4.3. Graphs

TABLE 4.3 Optimal Control Periodicity and Values of $K_a(\tau_{opt})$, $S(\tau^s_{opt})$, $C(\tau^c_{opt})$ for the System With Control Failures

Initial data						Results					
Kind of RV α distribution	$E\alpha$, h	$E\beta$, h	$E\gamma$, h	p_1	p_0	τ_{opt}, h	$K_a(\tau_{opt})$	τ^s_{opt}, h	$S(\tau^s_{opt})$ c.u./h	τ^c_{opt}, h	$C(\tau^c_{opt})$, c.u./h
Exponential	60	0.5	0.2	0	0	4.833	0.916	4.833	1.739	4.833	0.101
Exponential	60	0.5	0.2	0.3	0.25	5.094	0.867	5.522	1.568	48	1.292
Erlangian of the 2nd order	60	0.5	0.2	0.3	0.25	5.911	0.904	6.302	1.684	45.54	0.978
Erlangian of the 2nd order	60	0.5	0.2	0.2	0.25	5.007	0.903	5.259	1.688	36.11	0.817

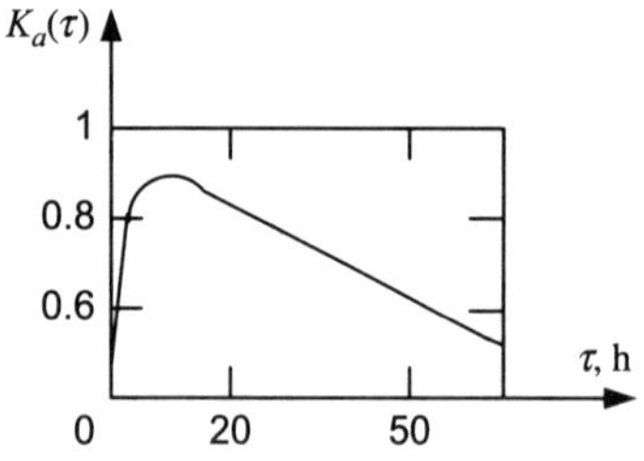

FIGURE 4.6 Graph of stationary availability factor $K_a(\tau)$ against control periodicity τ

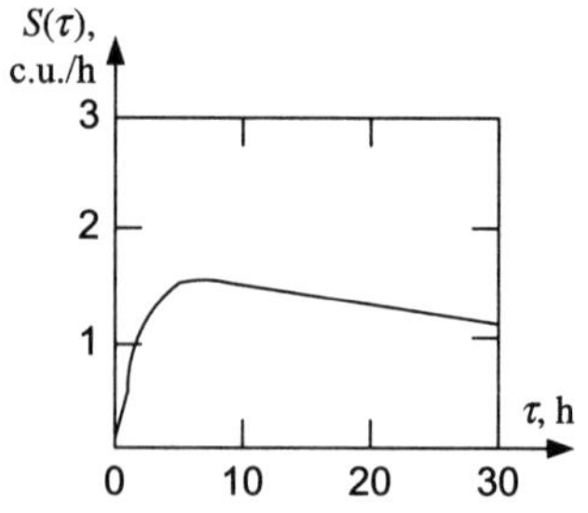

FIGURE 4.7 Graph of average specific income $S(\tau)$ against control periodicity τ

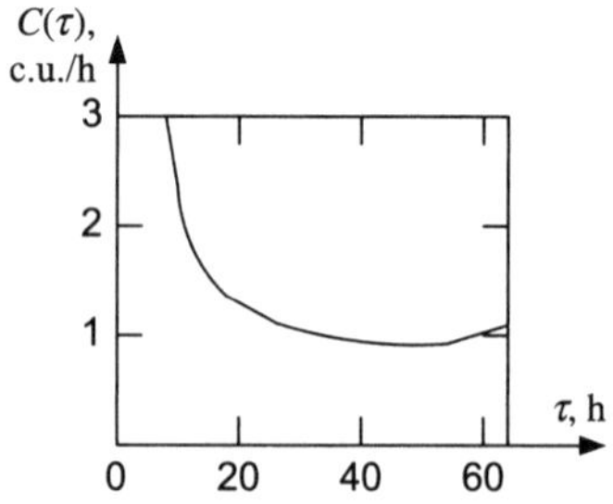

FIGURE 4.8 Graph of average specific expenses $C(\tau)$ against control periodicity τ

of $K_a(\tau)$, $S(\tau)$, $C(\tau)$, in the case of Erlangian distribution, under $p_1 = 0.3$, $p_0 = 0.25$, are represented in Figures 4.6, 4.7, and 4.8.

4.2 DEFINITION OF OPTIMAL CONTROL PERIODICITY FOR TWO-COMPONENT SYSTEMS

In contrast to one-component systems, we have to take into account not only control strategy but also reliability structure of the system, that is, parallel or serial component connection. Being objective functions, analytical expressions for system stationary characteristics, obtained in Chapter 3, make the foundation of control periodicity optimization a problem.

4.2.1 Control Periodicity Optimization for Two-Component Serial System

Applying formulas (3.49), (3.50), and (3.51) we can obtain optimal control periodicity of the system considered. To determine the optimal periods τ_{opt}, τ_{opt}^{s}, τ_{opt}^{c} ensuring maximum of stationary availability factor $K_a(\tau_{opt})$, maximum

TABLE 4.4 Optimal Control Periodicity and Values of $K_a(\tau_{opt})$, $S(\tau^s_{opt})$, $C(\tau^c_{opt})$

Initial data		Results					
$E\alpha_1$, h	$E\alpha_2$, h	τ_{opt}, h	$K_a(\tau_{opt})$,	τ^s_{opt}, h	$S(\tau^s_{opt})$, c.u./h	τ^c_{opt}, h	$C(\tau^c_{opt})$, c.u./h
90	70	3.096	0.924	3.541	4.384	4.355	0.24
90	50	2.793	0.915	3.198	4.321	3.927	0.267
90	10	1.459	0.845	1.686	3.76	2.041	0.533

of average specific income $S(\tau^s_{opt})$ per calendar time unit, and minimum of average specific expenses $C(\tau^c_{opt})$ per time unit of system up-state correspondingly, the following initial data are taken: average components time to failure $E\alpha_1, E\alpha_2$, average RT – $E\beta_1 = 0.100$ h, $E\beta_2 = 0.060$ h, average control duration $E\gamma = 0.125$ h, $c_1 = 5$ c.u., $c_2 = 4$ c.u., $c_3 = 3$ c.u., $c_4 = 2$ c.u. RV α_1, α_2 are assumed to have exponential distributions. The calculation results are given in Table 4.4. The graphs of functions $K_a(\tau)$, $S(\tau)$, $C(\tau)$, in case of exponential distribution, are represented in Figures 4.9, 4.10, and 4.11, respectively.

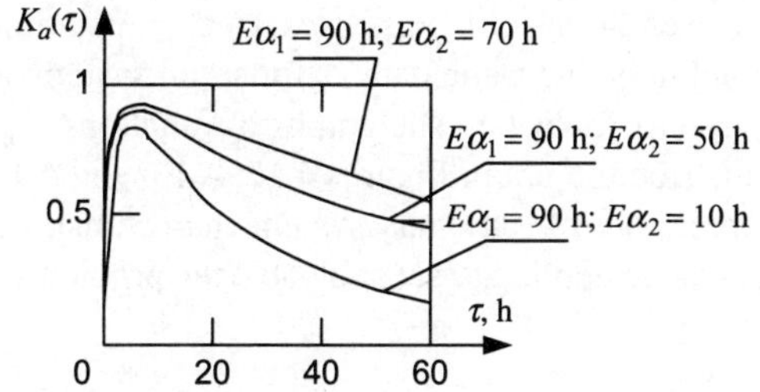

FIGURE 4.9 Graph of stationary availability factor $K_a(\tau)$ against control periodicity τ

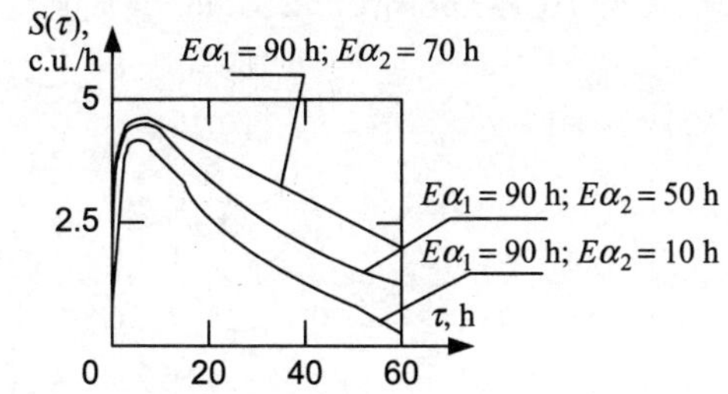

FIGURE 4.10 Graph of average specific income $S(\tau)$ against control periodicity τ

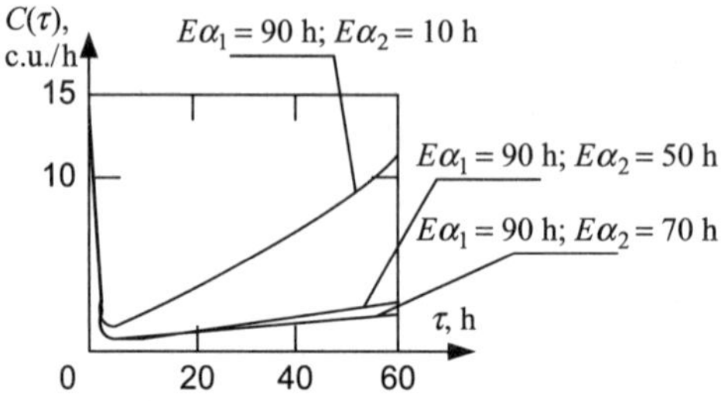

FIGURE 4.11 Graph of average specific expenses $C(\tau)$ against control periodicity τ

4.2.2 Control Periodicity Optimization for Two-Component Parallel System

Using expressions (3.153), (3.154), and (3.155) we define optimal control periodicity of the system. To calculate optimal values of latent failures control periodicity, the following inputs are taken: average components time to failure $E\alpha_1, E\alpha_2$; average restoration times $E\beta_1$, $E\beta_2$; average control duration $E\gamma$. RV α_1, α_2 are assumed to have exponential distributions. To obtain optimal periods τ_{opt}, τ_{opt}^s, τ_{opt}^c ensuring maximum of stationary availability factor $K_a(\tau_{opt})$, maximum of average specific income $S(\tau_{opt}^s)$ per calendar time unit, and minimum of average specific expenses $C(\tau_{opt}^c)$ per time unit of system up-state correspondingly, the same data as for serial system are used. Calculation results are put into Table 4.5. The graphs of functions $K_a(\tau)$, $S(\tau)$, $C(\tau)$ for exponential distribution are in Figures 4.12, 4.13, and 4.14, respectively.

Analytical expressions of stationary system characteristics, obtained in Sections 4.2 and 4.3, can be applied to set and solve the problems of multiobjective optimization.

TABLE 4.5 Optimal Control Periodicity and Values of $K_a(\tau_{opt})$, $S(\tau_{opt}^s)$, $C(\tau_{opt}^c)$

Initial data		Results					
$E\alpha_1$, h	$E\alpha_2$, h	τ_{opt}, h	$K_a(\tau_{opt})$	τ_{opt}^s, h	$S(\tau_{opt}^s)$, c.u./h	τ_{opt}^c, h	$C(\tau_{opt}^c)$, c.u./h
90	70	11.144	0.981	12.187	4.846	14.207	0.06
90	50	10.016	0.979	10.960	4.828	12.785	0.067
90	10	6.370	0.964	7.028	4.706	8.294	0.115

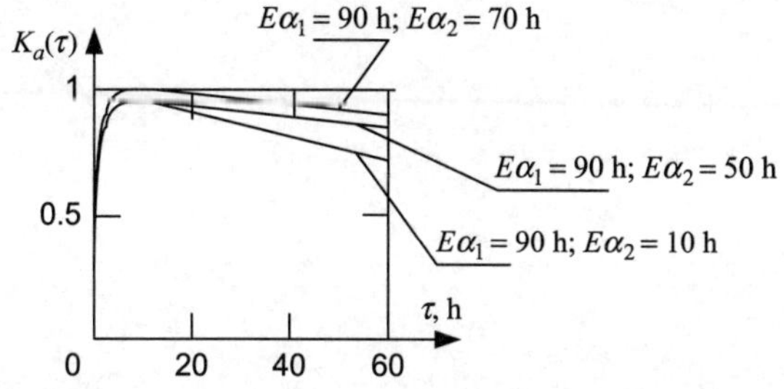

FIGURE 4.12 Graph of stationary availability factor $K_a(\tau)$ against control periodicity τ

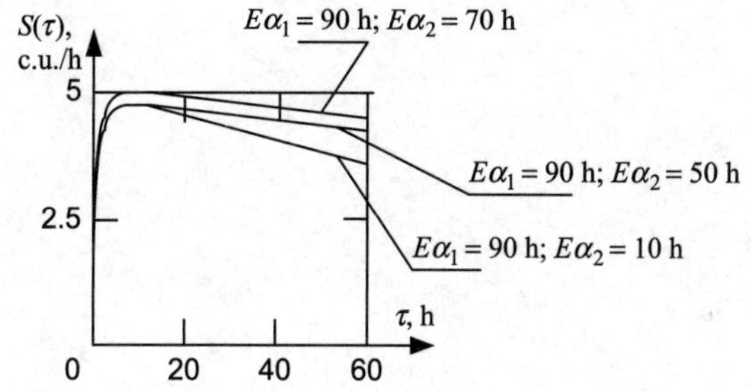

FIGURE 4.13 Graph of average specific income $S(\tau)$ against control periodicity τ

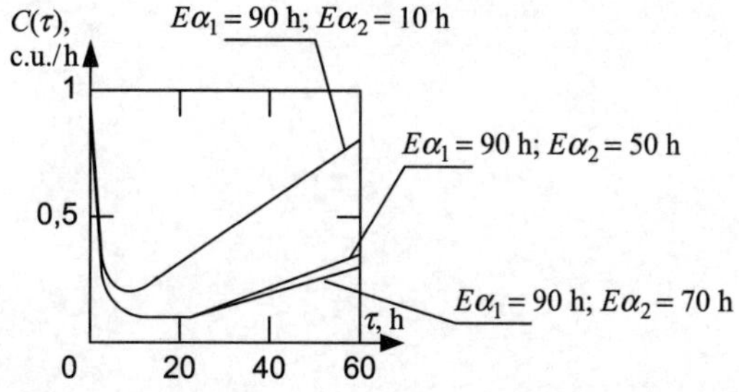

FIGURE 4.14 Graph of average specific expenses $C(\tau)$ against control periodicity τ

Chapter 5

Application and Verification of the Results

Chapter Outline

5.1 SIMULATION MODELS OF SYSTEMS WITH REGARD TO LATENT FAILURES CONTROL

Simulation modeling is efficient for systems investigation [24]. The method allows to gather necessary information on the system operation by making its virtual copy. The information is used for making system projects and research on the operation analysis.

In the present subsection, the comparison of the values of characteristics calculated by means of semi-Markov (SM) models in Chapters 2 and 3 with those obtained while simulation.

Constructing simulation models, we took the same assumptions as while analytical model building.

The language GPSS World [24] was applied for the simulation. The program for simulation consists of the blocks defining both parameters of model objects and actions performed over the objects.

Simulating one-component system with components deactivation, we admit all the RV (time to failure, control duration, and restoration time) to have Erlangian distributions of the 2nd order.

All the RV are assumed to have exponential distribution for the two-component system with component deactivation.

Semi-Markov Models. http://dx.doi.org/10.1016/B978-0-12-802212-2.00005-X
Copyright © 2015 Elsevier Inc. All rights reserved.

TABLE 5.1 Inputs for One-Component System with Deactivation

TF distribution	$E\alpha$, h	RT distribution	$E\beta$, h	$E\delta$, h
Erlangian of the 2nd order	100	Erlangian of the 2nd order	6	5

5.1.1 Comparison of Semi-Markov with Simulation Model in Case of One-Component System

Let us compare the values of stationary availability factor $K_a(\tau_{opt})$ and optimal control periodicity τ_{opt} for one-component system with component deactivation while control execution, obtained by means of formula (2.17) and by means of simulation technique, under the same inputs given in Table 5.1.

The results of analytical and simulation modeling, represented in Table 5.2, demonstrate the equality of optimal control periodicity for both models. It equals 20 h, while maximum values of stationary availability factor equal 0.7034 and 0.6900 correspondingly. Consequently, the difference of stationary availability factor is not more than 1.9%, which proves acceptable coincidence of results of analytical and simulation technique.

5.1.2 Comparison of Semi-Markov with Simulation Model in Case of Two-Component System

Let us compare the values of stationary availability factor $K_a(\tau_{opt})$ and optimal control periodicity τ_{opt} for two-component serial system, with component deactivation, obtained by means of formula (3.49) with those by simulation, under the same inputs represented in Table 5.3. Calculation results are placed in

TABLE 5.2 The Comparison of System Modeling Results

SM model (formula 2.17)		Simulation		
τ_{opt}, h	$K_a(\tau_{opt})$	τ_{opt}, h	$K_a(\tau_{opt})$	Error
20	0.7034	20	0.6900	1.9%

TABLE 5.3 Inputs for Double-Component System with Deactivation

$E\alpha_1$, h	$E\alpha_2$, h	$E\beta_1$, h	$E\beta_2$, h	$E\gamma$, h
90	70	0.1	0.067	0.125

TABLE 5.4 The Comparison of Modeling Results

SM model (formula 3.49)		Simulation		
τ_{opt}, h	$K_\alpha(\tau_{opt})$	τ_{opt}, h	$K_\alpha(\tau_{opt})$	Error
3.096	0.924	3.094	0.945	2.1%

Table 5.4. The comparison of analytical and simulation results shows the optimal control periodicity difference is 0.2%, while maximum values of stationary steady-state availability factor are 0.924 and 0.945, respectively. That means the difference of values of stationary availability factor is 2.1%.

The relative errors obtained for system stationary availability factor validate result of the analytical modeling.

5.2 THE STRUCTURE OF THE AUTOMATIC DECISION SYSTEM FOR THE MANAGEMENT OF PERIODICITY OF LATENT FAILURES CONTROL

One of the principal factors of engineering systems reliability and efficiency improvement is the creation of automatic decision systems for managing periodicity of control periodicity (ADS CPM). Automatic decision systems for control periodicity management ADS CPM, whose structure is proposed in the present section, allows to solve the following problem: to make the management schedule to optimize the process quality, according to the given data about the technological process. The maximum of stationary availability factor and average specific income per calendar time unit, the minimum of average specific expenses per time unit of system up state, and other criteria can be taken as objective ones.

One of the solutions of the given problem is ADS CPM of latent failures, performing management and information functions, and satisfying the operation principles: performance-informative, module, and hierarchic.

Performance-informative principle stands for control-restoration data mining, storage, analysis, and taking decision on control periodicity.

Module principle helps to make ADS CPM of latent failures of separate modules. It enables to correct information on the data obtained in the process of operation without essential expenses.

Hierarchic principle of ADS CPM of latent failures allows to coordinate management functions distribution among the structural levels: informational, strategic, tactic, and executive.

Informational level includes control data mining, storage, and processing. Strategic level implies the choice of RV distribution laws, system reliability structure and strategy of control execution. On tactic level, optimal reliability

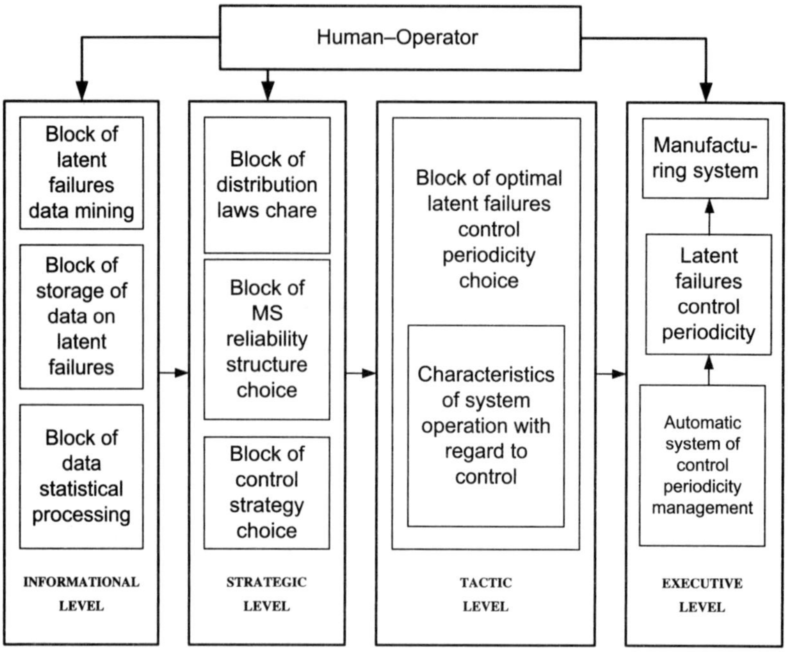

FIGURE 5.1 Hierarchic structure of ADS CPM of latent failures

and efficiency characteristics of manufacturing system (MS) with regard to control are determined. The executive level provides management of latent failures control periodicity by means of automatic management system. Herewith, a human – operator is a link, checking system operation.

The input is generated by a human-operator and taken from a database. The output includes the following information: optimal periodicity of latent failures control, extremal values of efficiency, and quality stationary characteristics.

Data input and operation control of ADS CPM of latent failures are made by a human-operator. The hierarchic structure of ADS CPM is given in Figure 5.1.

Structural scheme of ADS CPM of latent failures (Figure 5.2) is elaborated with regard to analogical systems. It is realized by means of program complex. The scheme consists of two basic modules: statistical and calculation one. Statistical module, in its turn, includes module of latent failures control process data mining and data processing. Module of data mining includes algorithms that allow to mine and store inputs of latent failures control. Module of data processing ensures statistical processing of inputs. Calculation module consists of algorithms of latent failures optimal control periodicity determination and of results visualization as graphs.

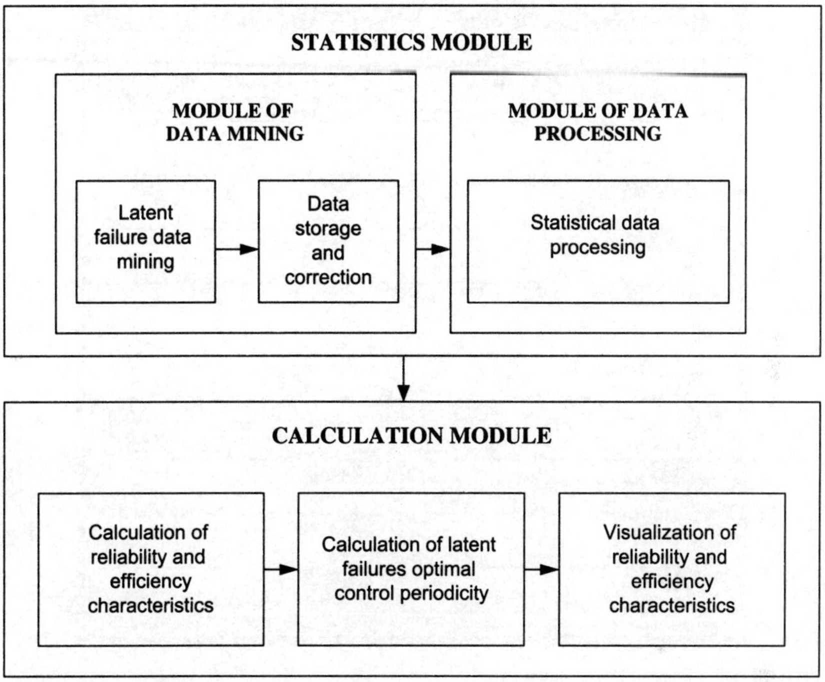

FIGURE 5.2 Structural scheme of ADS CPM of latent failures

5.2.1 Description of ADS CPM of Latent Failures Operation

The proposed ADS CPM of latent failures can be exploited by nonprofessionals. It has a multiwindow interface.

The ADS CPM software is designed for personal computers, compatible with IBM-PC/AT. The version of ADS CPM of latent failures is elaborated for personal computers with medium computational resources (Pentium, Pentium II, and their analogs). RAM capacity should be not less than 8 Mbytes, a processor – not worse than Pentium, a monitor SVGA with corresponding video adaptor, free disk memory capacity – not less than 10 Mbytes.

Windows 95 – Windows 7 can be the operating environment.

The high-level language Object Pascal was applied in integrated developer's environment Borland Delphi 6.0. The modules making the program are mutually independent. It allows to recurrently apply them for solving different programs.

Information on the number of system components, its reliability structure and control strategy is defined by the user of ADS CPM of latent failures. That is why the program has a dialogue module.

The interaction of program modules and multiwindow interface creation by means of Borland Delphi 6.0 are ensured by the basic window of ADS CPM of

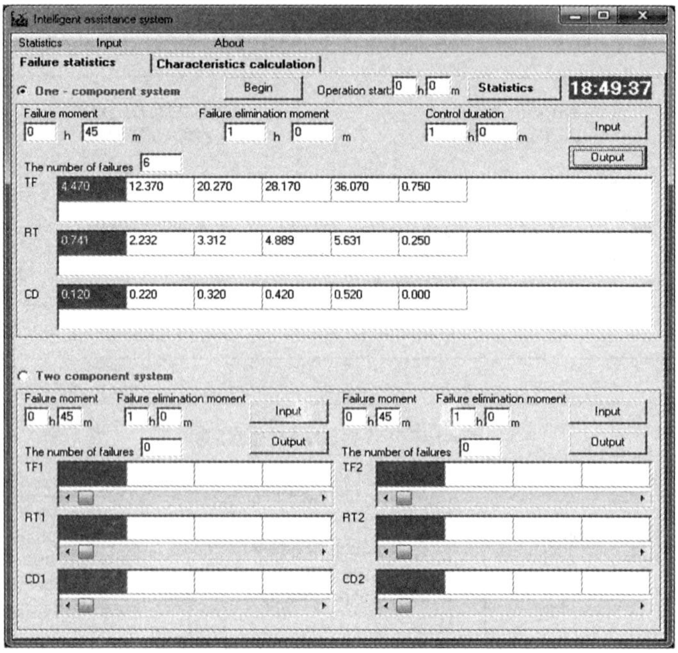

FIGURE 5.3 Page «Failure statistics»

latent failures. The basic window contains «Multi-page dialogue», which lets an operator to activate one of the pages: «Failure statistics» (Figure 5.3), «Statistical data processing» (Figure 5.4) or «Characteristics calculation» (Figure 5.5). The page «Failure statistics» ensures statistics compilation, in case distribution laws of time to failure, control duration, and restoration time are unknown a

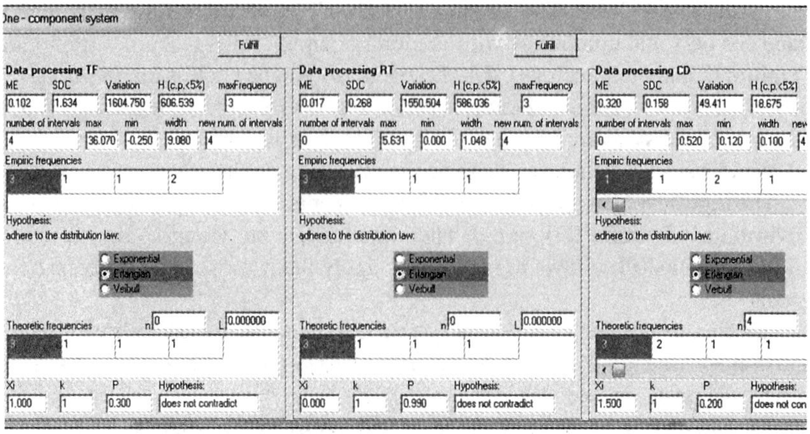

FIGURE 5.4 Page «Statistical data processing»

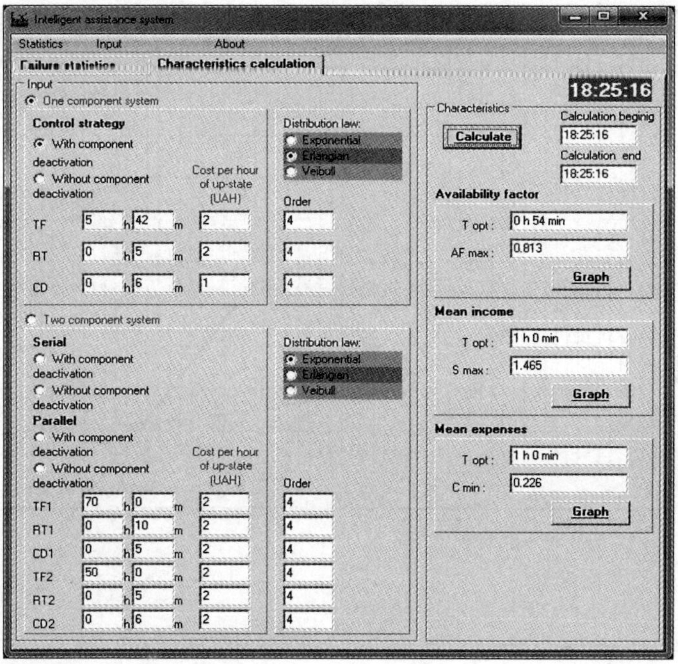

FIGURE 5.5 Page «Failure statistics»

priori. ADS CPM of latent failures saves the data and fixes the number of latent failures automatically.

On the page «Statistical data processing», such principal indexes of a statistical range and statistical characteristics as sample expectation, average square deviation are indicated. RV distribution laws are determined by means of data obtained.

The page «Characteristics calculation» enables to define the control strategy: either with operating component deactivation or without it. Keyboard input and automatic input from the pages «Statistical data input» and «Statistical data processing» are possible.

Pressing the button «Calculate» starts the module of calculation of maximum availability factor, maximum average specific income and minimum average specific expenses and definition of optimal control periodicity.

The button «Graph» corresponds results visualization module. The graphs of system characteristics are output. The example of the visualization module operation is given in Figure 5.6.

5.2.2 Passive Industrial Experiment

To validate mathematical models constructed in the present work, industrial experiment was carried out. The object of research is a miller with numeric program management. It is treated as a one-component system.

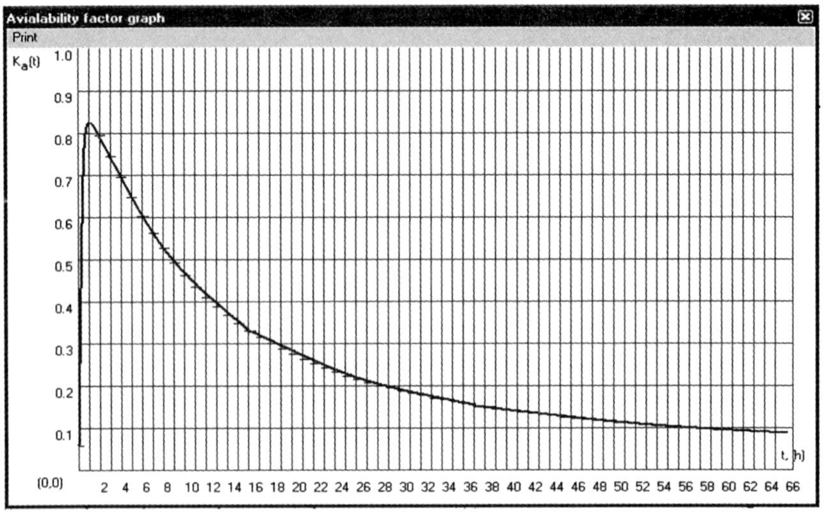

FIGURE 5.6 Graph of availability factor

The truncation of the miller cutting edge is discovered to be the most frequent failure. Cutting edge control is carried out each 2 h, according to the technological process. The experiment was to define the miller TF distribution law by means of control-measuring devices.

Next, semi-Markov models of control execution were verificated.

The chronometric data of miller time to failure are presented in Table 5.5. Table 5.6 includes the chronometric statistical range. In Tables 5.7 and 5.8, chronometric results of control failure and miller replacement are given.

In Table 5.9, experimental characteristics of miller reliability are depicted. Time to failure is concluded to obey Erlangian distribution of the 4th order.

χ^2 Pearson test for goodness of fit proved $\chi^2_{obser.} < \chi^2_{cr}$. It means theoretical distribution does not contradict experimental data obtained.

TABLE 5.5 Chronometric Results for Time to Failure, h

2.330	1.333	0.167	4.500	6.167	19.167	4.666	5.500	4.167
9.833	5.833	8.500	2.167	3.167	2.000	3.500	3.333	40.333
8.167	4.167	7.000	10.333	0.833	5.333	2.167	2.833	2.833
3.000	6.500	3.833	2.500	5.000	3.167	4.000	11.000	3.666
3.500	4.333	2.833	6.167	6.333	2.833	6.500	5.666	5.167
2.666	6.666	4.333	6.333	7.000	4.833	7.167	7.167	4.167

TABLE 5.6 Statistical Data Range of Cutter TF

	Interval	Interval midpoint	Frequency m_i	Theoretical frequency m'_i	Relative frequency m_i/n	Density $f_i(x)$	Density $f_{itheor}(x)$
1	0.167–2.164	1.1655	4	4	0.0741	0.0371	0.04
2	2.164–4.161	3.1625	18	19	0.3333	0.1686	0.1721
3	4.161–6.158	5.1595	16	16	0.2963	0.1484	0.1502
4	6.158–8.155	7.1565	9	8	0.1666	0.0834	0.0809
5	8.155–10.152	9.1535	3	4	0.0555	0.0278	0.0360
6	10.152–12.149	11.1505	2	2	0.0370	0.0185	0.0128
7	12.149–19.195	15.6720	1	1	0.0185	0.0026	0.0009
8	19.195–26.241	22.7180	0	0	0.0000	0.0000	1×10^{-5}
9	26.141–33.287	29.7640	0	0	0.0000	0.0000	7×10^{-9}
10	33.287–40.333	36.8100	1	0	0.0185	0.0026	3×10^{-12}

TABLE 5.7 Chronometric Results for Miller Control Duration, h

0.016	0.050	0.083	0.056	0.113	0.092	0.860	0.058	0.088
0.050	0.150	0.116	0.133	0.112	0.110	0.055	0.111	0.115
0.100	0.130	0.060	0.100	0.090	0.083	0.150	0.016	0.100
0.083	0.116	0.120	0.083	0.115	0.113	0.116	0.080	0.090
0.113	0.060	0.083	0.016	0.142	0.079	0.114	0.133	0.123
0.033	0.183	0.035	0.063	0.062	0.110	0.072	0.115	0.113

TABLE 5.8 Chronometric Results for Miller Restoration, h

0.058	0.088	0.076	0.077	0.075	0.082	0.091	0.089	0.090
0.080	0.090	0.078	0.060	0.090	0.086	0.089	0.076	0.077
0.086	0.079	0.079	0.080	0.092	0.079	0.080	0.081	0.091
0.090	0.076	0.077	0.125	0.081	0.091	0.076	0.092	0.075
0.076	0.091	0.082	0.088	0.088	0.075	0.079	0.077	0.080
0.075	0.075	0.092	0.084	0.079	0.080	0.09	0.075	0.091

TABLE 5.9 Experimental Characteristics of the Miller Reliability with Regard to Latent Failures Control

Reliability characteristics	Expectation $M^*(X)$, h	Variance σ^{*2}, h^2	Variation coefficient V,%	Order of Erlangian distribution	Pearson test for goodness of fit χ^2_{cr}	Pearson test for goodness of fit $\chi^2_{Hab\pi}$	Stationary availability factor/ opt. Control period, h
Time to failure	5.700	25.640	88.830	4	20.100	0.428	0.813/ 0.968
Control duration	0.100	–	–	–			
Restoration time	0.083	–	–	–			

TABLE 5.10 Passive Experiment Results

	Experimental model	Semi-Markov model	
System	$K_\alpha(\tau = 2\,\text{h})$	$K_\alpha(\tau = 2\,\text{h})$	Error
One-component with deactivation	0.759	0.798	3.9%

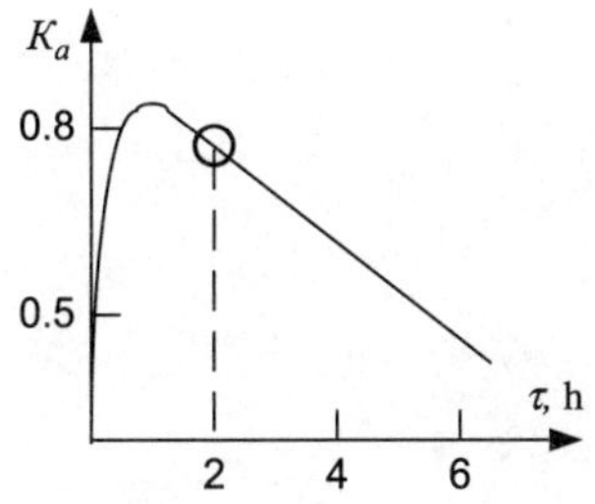

FIGURE 5.7 Graph of stationary availability factor $K_a(\tau)$ against control periodicity τ

System stationary availability factor is calculated on the basis of data obtained. In Table 5.10, comparative results of theoretical and experimental data are given.

Table 5.10 shows calculation results do not differ from experimental ones for more than 4%. It confirms sufficient level of the validity of one-component model constructed in Section 2.

Verification by means of ADS CPM of latent failures results in the optimal control periodicity $\tau_{opt} = 0.9$ h and corresponding availability factor $K_a = 0.813$. Graph of stationary availability factor against control periodicity τ (Figure 5.7) vividly demonstrates periodicity $\tau = 2$ h applied in industry not to be optimal. The value of control periodicity improves stationary availability factor for 1.6%.

Chapter 6

Semi-Markov Models of Systems of Different Function

Chapter Outline

6.1 SEMI-MARKOV MODEL OF A QUEUING SYSTEM WITH LOSSES

6.1.1 System Description

In this section, semi-Markov model of $GI/\vec{G}/N/0$ queuing system with losses (QS) is built. Stationary probabilistic and time $M/\vec{G}/N/0$ QS characteristics are obtained [12, 25].

The loss QS with N servers is considered. Let request service time be RV α_k with DF $F_k(x)$ and DD $f_k(x)$, for k^{th} server, $k=\overline{1,N}$; interarrival time is RV β with DF $G(x)$ and DD $g(x)$. The arriving request is directed to one of free servers with the same probability. It is lost if there are no free servers. RVs α_k, β are assumed to be independent and to have finite expectations.

Semi-Markov Models. http://dx.doi.org/10.1016/B978-0-12-802212-2.00018-8
Copyright © 2015 Elsevier Inc. All rights reserved.

6.1.2 Semi-Markov Model Building

To describe the QS $GI/\vec{G}/N/0$ operation, let us introduce the following set E of semi-Markov system states:

$$E=\{i\bar{d}\bar{x}: i=\overline{1,N},\quad \bar{d}=(d_1,\dots,d_k,\dots,d_N),\quad \bar{x}=(x_1,\dots,x_k,\dots,x_N),$$

$$x_k\ge 0,\ k=\overline{1,N}\}\,U\{i\bar{d}\bar{x}z: i=\overline{1,N},\quad \bar{d}=(d_1,\dots,d_k,\dots,d_N),$$

$$\bar{x}=(x_1,\dots,x_k,\dots,x_N),\quad x_k\ge 0,\ k=\overline{1,N},\quad z\ge 0\}U$$

$$U\{i\bar{x}: i=\overline{1,N},\quad \bar{x}=(x_1,\dots,x_k,\dots,x_N),\quad x_i\ge 0,\quad x_k\ge x_i,\ k=\overline{1,N}\},$$

where

(a) states $i\bar{d}\bar{x}$ define system at the moments of request arrival if there are free servers; the integer i indicates the number of a server that has begun to serve a request; the vector $\bar{d}$ defines the state of servers:

$$d_k=\begin{cases}0, & \text{if the k}^{\text{th}}\text{ server is free},\\ 1, & \text{if the k}^{\text{th}}\text{ server is busy},\end{cases}$$

x_k is time elapsed since the beginning of request service by the kth server, $d_i=1$, $x_i=0$, $x_k=0$ if $d_k=0$;

(b) the states $i\bar{d}\bar{x}z$ describe the system at the moment of the ith server emptying; z is time elapsed since the moment of last request arrival. Vectors $\bar{x}$ and $\bar{d}$ have the above interpretation;

(c) the state $i\bar{x}$ means the request is lost due to the absence of free servers; i is the number of the last busy server. The components of the vector $\bar{x}$ preserve the same meaning.

Let us construct SMP $\xi(t)$ with the phase space of states E. The process describes the $GI/\vec{G}/N/0$ QS operation. We introduce the MRP $\{\xi_n,\theta_n;\ n\ge 0\}$ to determine the process. Let us introduce the following indications: let $\bar{d}=(d_1,\dots,d_k,\dots,d_N)$, then

$$d'_k=\begin{cases}0, \text{ if } d_k=1,\\ 1, \text{ if } d_k=0,\end{cases}\qquad (\bar{d}^{(i)})_k=\begin{cases}d_k, & k\ne i,\\ d'_i, & k=i;\end{cases}$$

$|\bar{d}|$ is the number of vector $\bar{d}$ components, which are equal to 0 (the number of free servers); if $\bar{x}=(x_1,\dots,x_k,\dots x_N)$, the vectors $(\bar{x}+t)^{(\bar{d})}$, $((\bar{x}+t)_i^{(\bar{d})},t_i)$, $t\in R$, $t_i\ge 0$ are defined in the following way:

$$[(\bar{x}+t)^{(\bar{d})}]_k=\begin{cases}x_k+t, & \text{if } d_k=1,\\ 0, & \text{if } d_k=0,\end{cases}$$

$$[((\bar{x}+t)_i^{(\bar{d})},\ t_i)]_k = \begin{cases} x_k + t, & \text{if } d_k = 1, k \neq i, \\ 0, & \text{if } d_k = 0,\ k \neq i, \\ t_i, & \text{if } k = i. \end{cases}$$

Let us describe the probabilities of EMC $\{\xi_n; n \geq 0\}$ transitions. Let us consider transitions from the states $i\bar{d}\bar{x}z$. For the states $i\bar{d}\bar{x}$, $i\bar{x}$ it can be done in the same way.

1. Consider the state $i\bar{d}\bar{x}z$ $|\bar{d}| = N$ (all the servers are free; the state is $i\bar{0}\bar{0}z$). In this case, transitions to the states $j\bar{0}^{(j)}\bar{0}$, $j = \overline{1,N}$ are possible. The transition probability is calculated by the formula:

$$P_{i\bar{0}\bar{0}z}^{j\bar{0}^{(j)}\bar{0}} = 1/N. \tag{6.1}$$

2. Introduce transitions from the states $i\bar{d}\bar{x}z$, $1 \leq |\bar{d}| \leq N-1$:

(a) $i\bar{d}\bar{x}z \to i\bar{d}^{(j)}\bar{y}$ under the restrictions: $j: d_j = 0$, $|\bar{d}^{(j)}| = |\bar{d}| - 1$, $\bar{y} = (\bar{x}+t)^{(\bar{d})}$, $0 \leq t < \infty$,

$$p_{i\bar{d}\bar{x}z}^{j\bar{d}^{(j)}\bar{y}} = \frac{1}{|\bar{d}|} \frac{g(z+t) \prod\limits_{k:d_k=1} \bar{F}_k(x_k+t)}{\prod\limits_{k:d_k=1} \bar{F}_k(x_k)\bar{G}(z)}, \tag{6.2}$$

(b) $i\bar{d}\bar{x}z \to j\bar{d}^{(j)}\bar{y}z_1$ under the conditions: $d_j = 1$, $|\bar{d}^{(j)}| = |\bar{d}| + 1$, $\bar{y} = (\bar{x}+t)^{(\bar{d}^{(j)})}$, $z_1 = z + t$, $0 \leq t < \infty$.
The probability density of the transition is:

$$p_{i\bar{d}\bar{x}z}^{j\bar{d}^{(j)}\bar{y}z_1} = \frac{f_j(x_j+t) \prod\limits_{\substack{k:d_k=1,\\ k\neq j}} \bar{F}_k(x_k+t)\bar{G}(z+t)}{\prod\limits_{k:d_k=1} \bar{F}_k(x_k)\bar{G}(z)}. \tag{6.3}$$

Let us write out the distributions of sojourn times:

$$\mathrm{P}(\theta_{i\bar{d}\bar{x}} \geq t) = \prod_{k:d_k=1} \frac{\bar{F}_k(x_k+t)}{\bar{F}_k(x_k)} \bar{G}(t),$$

$$\mathrm{P}(\theta_{i\bar{x}} \geq t) = \prod_{k=1}^{N} \frac{\bar{F}_k(x_k+t)}{\bar{F}_k(x_k)} \bar{G}(t), \tag{6.4}$$

$$\mathrm{P}(\theta_{i\bar{d}\bar{x}z} \geq t) = \prod_{k:d_k=1} \frac{\bar{F}_k(x_k+t)}{\bar{F}_k(x_k)} \frac{\bar{G}(z+t)}{\bar{G}(z)}.$$

6.1.3 EMC Stationary Distribution Determination

Let us apply the $GI/\vec{G}/N/0$ QS semi-Markov model to obtain the $M/\vec{G}/N/0$ QS stationary characteristics. The exponential distribution of the input flow is supposed: $G(x)=1-e^{-bx}$, $x\geq 0$.

Let $\rho(i\bar{d}\bar{x})$, $0\leq|\bar{d}|\leq N-1$, $\rho(i\bar{x})$ be the densities of the EMC $\{\xi_n;\ n\geq 0\}$ stationary distribution. Introduce the indications:

$$\tilde{\rho}(i\bar{d}\bar{x})=\frac{\rho(i\bar{d}\bar{x})}{\prod\limits_{k:d_k=1}\bar{F}_k(x_k)},\tilde{\rho}(i\bar{x})=\frac{\rho(i\bar{x})}{\prod\limits_{k=1}^{N}\bar{F}_k(x_k)}. \tag{6.5}$$

With the help of probabilities and probability densities of the EMC $\{\xi_n;\ n\geq 0\}$ transition probabilities (6.1)–(6.3) and of the formula (1.15), let us construct the system of integral equations for the functions $\tilde{\rho}(i\bar{d}\bar{x})$, $\tilde{\rho}(i\bar{x})$, $\tilde{\rho}(i\bar{0}\bar{0})$, $\tilde{\rho}(i\bar{0}^{(i)}\bar{0})$:

$$\tilde{\rho}(i\bar{x})=g(x_i)\tilde{\rho}(i\bar{1}(\bar{x}-x_i)^{(\bar{1})})+\int_0^{x_i}\tilde{\rho}(i(\bar{x}-t)^{(\bar{1})})g(t)\,dt; \tag{6.6}$$

$$\tilde{\rho}(i\bar{d}\bar{x})=\frac{1}{(|\bar{d}|+1)}g(x_j)\tilde{\rho}(i\bar{d}^{(i)}(\bar{x}-x_j)^{(\bar{d}^{(i)})})+$$

$$+\frac{1}{(|\bar{d}|+1)}\sum_{\substack{m:d_m=0,\\ m\neq i}}\int_0^{x_j}g(t)\tilde{\rho}(m\bar{d}^{(i)}(\bar{x}-t)^{(\bar{d}^{(i)})})\,dt,x_j=\bigwedge_{\substack{k:d_k=1,\\ k\neq i}}x_k, \tag{6.7}$$

in the case of states $i\bar{d}\bar{x}$: $0\leq|\bar{d}|\leq N-2$, $d_i=1$, Λ is a minimum sign;

$$\tilde{\rho}(i\bar{0}^{(i)}\bar{0})=\frac{1}{N}\sum_{m=1}^{N}\tilde{\rho}(m\bar{0}\bar{0}); \tag{6.8}$$

$$\tilde{\rho}(i\bar{1}^{(i)}\bar{x})=\int_0^{\infty}\bar{G}(x_j)f_i(t_i+x_j)\tilde{\rho}(j\bar{1}((\bar{x}-x_j)_i,t_i))dt_i+$$

$$+\int_0^{x_j}f_i(t)\bar{G}(t)\tilde{\rho}(i\bar{1}(\bar{x}-t)^{(\bar{1}^{(i)})})dt+$$

$$+\int_0^{x_j}dt\int_{x_j}^{\infty}f_i(t+t_i)\bar{G}(t)\tilde{\rho}(i((\bar{x}-t)_i,t_i))dt_i+$$

$$+\int_0^{x_j}dt\int_0^{x_j-t}f_i(t+t_i)\bar{G}(t)\tilde{\rho}(i((\bar{x}-t)_i,t_i))dt_i,x_j=\bigwedge_{k:k\neq i}x_k; \tag{6.9}$$

$$\tilde{\rho}(i\bar{d}\bar{x}) = \int_0^{x_j} f_i(x_j + t_i)\bar{G}(x_j)\tilde{\rho}(i\bar{d}^{(i)}((\bar{x} - x_j))_i^{(\bar{d})}, t_i))\,dt_i +$$

$$+ \sum_{\substack{m:d_m=0,\\ m\neq i}} \int_0^{\infty} dt_i \int_0^{x_j} f_i(t_i + t)\bar{G}(t)\tilde{\rho}(m\bar{d}^{(i)}((\bar{x} - t)_i^{(\bar{d})}, t_i))\,dt +$$

$$+ \int_0^{x_j} f_i(t)\bar{G}(t)\tilde{\rho}(i\bar{d}^{(i)}(\bar{x} - t)^{(\bar{d})})\,dt, x_j = \bigwedge_{k:d_k=1} x_k; \tag{6.10}$$

for the states $i\bar{d}\bar{x}$: $2 \le |\bar{d}| \le N-1, \quad d_i = 0;$

$$\tilde{\rho}(i\bar{0}\bar{0}) = P\{\beta > \alpha_i\}\tilde{\rho}(i\bar{0}^{(i)}\bar{0}) +$$

$$+ \sum_{\substack{m:d_m=0,\\ m\neq i}} \int_0^{\infty} dx_i \int_0^{\infty} f_i(x_i + t)\bar{G}(t)\tilde{\rho}(m\bar{0}^{(i)}(0,...,0,x_i,0,...,0))\,dt; \tag{6.11}$$

$$\sum_{i=1}^{N} [\,\tilde{\rho}(i\bar{0}\bar{0}) + \tilde{\rho}(i\bar{0}^{(i)}\bar{0}) + \int_0^{\infty} dx_i \int_{x_i}^{\infty} \cdots \int_{x_i}^{\infty} \tilde{\rho}(i\bar{x}) \prod_{k=1}^{N} \bar{F}_k(x_k)dx_1 ...$$

$$...dx_{i-1}dx_{i+1}...dx_N\,] + \sum_{\substack{i\bar{d}:\\ i\bar{d}\neq i\bar{0},\, i\bar{0}^{(i)}}} \int_{R_+^{(i\bar{d})}} \rho(i\bar{d}\bar{x}) \prod_{k:d_k=1} \bar{F}_k(x_k)d\bar{x}^{(i\bar{d})} = 1. \tag{6.12}$$

Lemma. The system of equations (6.6)–(6.12) has the following solution:

$$\tilde{\rho}(i\bar{x}) = \rho_0 b;\ \tilde{\rho}(i\bar{d}\bar{x}) = \rho_0 \frac{|\bar{d}|!}{b^{|\bar{d}|}} \text{ if } d_i = 1;$$

$$\tilde{\rho}(i\bar{d}\bar{x}) = \rho_0 \frac{(|\bar{d}|-1)!}{b^{|\bar{d}|-1}} \text{ if } d_i = 0, \tag{6.13}$$

where $\rho_0 = \left(\sum_{i\bar{d}:d_i=1} \frac{|\bar{d}|!}{b^{|\bar{d}|}} \prod_{\substack{k:d_k=1,\\ k\neq i}} E\alpha_k + \sum_{i\bar{d}:d_i=0} \frac{(|\bar{d}|-1)!}{b^{|\bar{d}|-1}} \prod_{k:d_k=1} E\alpha_k + b\prod_{k=1}^{N} E\alpha_k \right)^{-1}.$

Lemma proof is carried out by means of the direct substitution of functions (6.13) into the system of equations (6.6)–(6.11); the constant ρ_0 is obtained from the normalization requirement (6.12).

From the lemma we conclude that the EMC $\{\xi_n; n \ge 0\}$ stationary distribution is:

$$\rho(i\bar{x}) = \rho_0 b \prod_{k=1}^{N} \bar{F}_k(x_k);\ \rho(i\bar{d}\bar{x}) = \rho_0 \frac{|\bar{d}|!}{b^{|\bar{d}|}} \prod_{k:d_k=1} \bar{F}_k(x_k), d_i = 1;$$

$$\rho(i\bar{d}\bar{x}) = \rho_0 \frac{(|\bar{d}|-1)!}{b^{|\bar{d}|-1}} \prod_{k:d_k=1} \bar{F}_k(x_k), d_i = 0. \tag{6.14}$$

6.1.4 System Stationary Characteristics Determination

Let us formulate the theorem to obtain stationary characteristics of the $M/\vec{G}/N/0$ QS.

Theorem [18]. Let the embedded Markov chain into the SMP $\xi(t)$ with phase state (E, B) be a recurrent one with invariant measure ρ and $\int_E m(x)\rho(dx) < \infty$. Then, if the process $\xi(t)$ is nonlattice, then

$$\lim_{t\to\infty} P_y\{\xi(t) \in A\} = \frac{\int_A m(x)\rho(dx)}{\int_E m(x)\rho(dx)}, y \in E, A \in B, \tag{6.15}$$

where $m(x)$ is the average time of the process $\xi(t)$ sojourn time in the state $x \in E$.

Let us determine the averages of sojourn times in the states of SMP $\xi(t)$ describing the $M/\vec{G}/N/0$ QS operation by using (6.4). We get:

$$E\theta_{i\bar{d}\bar{x}} = \int_0^\infty \frac{\prod_{k:d_k=1} \bar{F}_k(x_k+t)\bar{G}(t)}{\prod_{k:d_k=1} \bar{F}_k(x_k)} dt,$$

$$E\theta_{\bar{i}\bar{x}} = \int_0^\infty \frac{\prod_{k=1}^{N} \bar{F}_k(x_k+t)\bar{G}(t)}{\prod_{k=1}^{N} \bar{F}_k(x_k)} dt. \tag{6.16}$$

Introduce the notation D as the set of all the binary vectors $\bar{d}$ describing the states of system servers. And, let D_0 be some subset of D, and $\tilde{D}_0 = \{i\bar{d}\bar{x} : \bar{d} \in D_0\}$ or $\{i\bar{d}\bar{x} : \bar{d} \in D_0\} \cup \{\bar{i}\bar{x}\}$, if $\bar{1} \in D_0$. By means of $\tilde{D}_0$ we can determine the stationary probability that the QS is in states belonging to the subset D_0, that is, we calculate $\lim_{t\to\infty} P_e\{\xi(t) \in \tilde{D}_0\}$, $e \in E$ applying the formula (6.15).

Applying (6.14), (6.16) we get:

$$\int_{\tilde{D}_0} m(e)\rho(de) = \rho_0 \sum_{\bar{d}\in D_0} \Bigg[\frac{|\bar{d}|!}{b^{|\bar{d}|}} \sum_{i:d_i=1} \underbrace{\int_0^\infty \cdots \int_0^\infty}_{N-|\bar{d}|-1} \int_0^\infty \prod_{\substack{k:d_k=1,\\ k\neq i}} \bar{F}_k(x_k+t)\bar{F}_i(t)\bar{G}(t)dtdx_{k_1}\ldots$$

$$\ldots dx_{k_{N-|\bar{d}|-1}} + \frac{(|\bar{d}|-1)!}{b^{|\bar{d}|-1}} \sum_{i:d_i=0} \underbrace{\int_0^\infty \cdots \int_0^\infty}_{N-|\bar{d}|} \int_0^\infty \prod_{k:d_k=1} \bar{F}_k(x_k+t)\bar{G}(t)\, dtdx_{k_1}\ldots dx_{k_{N-|\bar{d}|}} \Bigg] =$$

$$= \rho_0 \sum_{\bar{d} \in D_0} \left[\frac{|\bar{d}|!}{b^{|\bar{d}|}} \sum_{i:d_i=1} \underbrace{\int_0^\infty \cdots \int_0^\infty}_{N-|\bar{d}|-1} \int_0^\infty \prod_{\substack{k:d_k=1,\\ k\neq i}} \bar{F}_k(x_k+t)\bar{F}_i(t)\bar{G}(t)\, dt dx_{k_1} \ldots \right. \tag{6.17}$$

$$\left. \ldots dx_{k_{N-|\bar{d}|-1}} + \frac{|\bar{d}|!}{b^{|\bar{d}|-1}} \underbrace{\int_0^\infty \cdots \int_0^\infty}_{N-|\bar{d}|} \int_0^\infty \prod_{k:d_k=1} \bar{F}_k(x_k+t)\,\bar{G}(t)\, dt dx_{k_1} \ldots dx_{k_{N-|\bar{d}|}} \right].$$

Using the formula of integration by parts in the second summand, we get:

$$\underbrace{\int_0^\infty \cdots \int_0^\infty}_{N-|\bar{d}|} \int_0^\infty \prod_{k:d_k=1} \bar{F}_k(x_k+t)\,\bar{G}(t)\, dt dx_{k_1} \ldots dx_{k_{N-|\bar{d}|}} = \frac{1}{b} \prod_{k:d_k=1} E\alpha_k -$$

$$-\frac{1}{b} \sum_{m=1}^{N-|\bar{d}|} \underbrace{\int_0^\infty \cdots \int_0^\infty}_{N-|\bar{d}|-1} \int_0^\infty \prod_{\substack{k:d_k=1,\\ k\neq m}} \bar{F}_k(x_k+t)\bar{F}_m(t)\,\bar{G}(t)\, dt dx_{k_1} \ldots dx_{k_{m-1}} dx_{k_{m+1}} \ldots dx_{k_{N-|\bar{d}|}}.$$

By substituting this expression into (6.17) and canceling the summands we have:

$$\int_{\tilde{D}_0} m(e)\rho(de) = \rho_0 \sum_{\bar{d} \in D_0} \frac{|\bar{d}|!}{b^{|\bar{d}|}} \prod_{k:d_k=1} E\alpha_k. \tag{6.18}$$

Thus, relying on the equality (6.15) we get:

$$P(D_0) = \lim_{t\to\infty} P_e\{\xi(t) \in \tilde{D}_0\} = \frac{\sum_{\bar{d}\in D_0} |\bar{d}|! E^{|\bar{d}|}\beta \prod_{k:d_k=1} E\alpha_k}{\sum_{\bar{d}\in D} |\bar{d}|! E^{|\bar{d}|}\beta \prod_{k:d_k=1} E\alpha_k}. \tag{6.19}$$

The formula (6.19) can be given as follows:

$$P(D_0) = \frac{\sum_{\bar{d}\in D_0} |\bar{d}|! E^{|\bar{d}|}\beta \prod_{k:d_k=1} E\alpha_k}{\sum_{m=0}^{N} (N-m)! E^{N-m}\beta \sum_{\{l_1,\ldots,l_m\}} E\alpha_{l_1} \cdot \ldots \cdot E\alpha_{l_m}}, \tag{6.20}$$

where $\{l_1,\ldots,l_m\}$ is some combination of the set $\{1,\ldots,N\}$ taken m at a time.

According to the formula (6.20), the stationary probability that k servers are busy is

$$P_k = \frac{(N-k)!\, E^{N-k}\beta \sum_{\{l_1,\ldots,l_k\}} E\alpha_{l_1} \cdot \ldots \cdot E\alpha_{l_k}}{\sum_{m=0}^{N} (N-m)!\, E^{N-m}\beta \sum_{\{l_1,\ldots,l_m\}} E\alpha_{l_1} \cdot \ldots \cdot E\alpha_{l_m}}. \tag{6.21}$$

Let us represent the set D as the union $D = D_+ \bigcup D_-$, $D_+ \bigcap D_- = \varnothing$, and let $\tilde{D}_+, \tilde{D}_-$ be the subsets of E, which correspond to D_+, D_- and are defined in the same way as the set $\tilde{D}_0$.

Let us determine the SMP $\xi(t)$ average stationary sojourn time T_{D_+} in the subset $\tilde{D}_+$. We shall apply the formula (1.17):

$$\int_{\tilde{D}_+} P(e,\tilde{D}_-)\,\rho\,(de) =$$

$$= \rho_0 \sum_{\bar{d}\in D_+} \left\{ \sum_{i:d_i=1} \left[\sum_{\substack{j:d_j=1,\\ \bar{d}^{(j)}\in D_-}} \underbrace{\int_0^\infty \ldots \int_0^\infty}_{N-|\bar{d}|-1} \int_0^\infty \frac{|\bar{d}\,|!}{b^{|\bar{d}|}} f_j(x_j+t) \prod_{\substack{k:d_k=1,\\ k\neq j}} \bar{F}_k(x_k+t)\,\bar{G}\,(t)\,dtdx_{k_1}\ldots dx_{k_{N-|\bar{d}|}} + \right.\right.$$

$$\left. + \sum_{\substack{j:d_j=0,\\ \bar{d}^{(j)}\in D_-}} \underbrace{\int_0^\infty \ldots \int_0^\infty}_{N-|\bar{d}|-1} \int_0^\infty \frac{|\bar{d}\,|!}{b^{|\bar{d}|}} \frac{1}{|\bar{d}\,|} g(t) \prod_{k:d_k=1} \bar{F}_k(x_k+t)\,dtdx_{k_1}\ldots dx_{k_{N-|\bar{d}|}} \right] +$$

$$+ \sum_{i:d_i=0} \left[\sum_{\substack{j:d_j=1,\\ \bar{d}^{(j)}\in D_-}} \underbrace{\int_0^\infty \ldots \int_0^\infty}_{N-|\bar{d}|} \int_0^\infty \frac{(|\bar{d}\,|-1)!}{b^{|\bar{d}|-1}} f_j(x_j+t) \prod_{\substack{k:d_k=1,\\ k\neq j}} \bar{F}_k(x_k+t)\,\bar{G}\,(t)\,dtdx_{k_1}\ldots dx_{k_{N-|\bar{d}|}} + \right.$$

$$\left.\left. + \sum_{\substack{j:d_j=0,\\ \bar{d}^{(j)}\in D_-}} \underbrace{\int_0^\infty \ldots \int_0^\infty}_{N-|\bar{d}|} \int_0^\infty \frac{(|\bar{d}\,|-1)!}{b^{|\bar{d}|-1}} \frac{1}{|\bar{d}\,|} g(t) \prod_{k:d_k=1} \bar{F}_k(x_k+t)\,dtdx_{k_1}\ldots dx_{k_{N-|\bar{d}|}} \right]\right\} =$$

$$= \rho_0 \sum_{\bar{d}\in D_+} \left\{ \sum_{i:d_i=1} \left[\sum_{\substack{j:d_j=1,\\ \bar{d}^{(j)}\in D_-}} \underbrace{\int_0^\infty \ldots \int_0^\infty}_{N-|\bar{d}|-1} \int_0^\infty \frac{|\bar{d}\,|!}{b^{|\bar{d}|}} f_j(x_j+t) \prod_{\substack{k:d_k=1,\\ k\neq j}} \bar{F}_k(x_k+t)\,\bar{G}\,(t)\,dtdx_{k_1}\ldots dx_{k_{N-|\bar{d}|}} + \right.\right.$$

$$\left. + \sum_{\substack{j:d_j=0,\\ \bar{d}^{(j)}\in D_-}} \underbrace{\int_0^\infty \ldots \int_0^\infty}_{N-|\bar{d}|-1} \int_0^\infty \frac{(|\bar{d}\,|-1)!}{b^{|\bar{d}|}} g(t) \prod_{k:d_k=1} \bar{F}_k(x_k+t)\,dtdx_{k_1}\ldots dx_{k_{N-|\bar{d}|}} \right] +$$

$$+ \sum_{\substack{j:d_j=1,\\ \bar{d}^{(j)}\in D_-}} \underbrace{\int_0^\infty \ldots \int_0^\infty}_{N-|\bar{d}|-1} \int_0^\infty \frac{|\bar{d}\,|!}{b^{|\bar{d}|-1}} \bar{F}_j(t) \prod_{\substack{k:d_k=1,\\ k\neq j}} \bar{F}_k(x_k+t)\,\bar{G}\,(t)\,dtdx_{k_1}\ldots dx_{k_{N-|\bar{d}|}} +$$

$$\left. + \sum_{\substack{j:d_j=0,\\ \bar{d}^{(j)}\in D_-}} \underbrace{\int_0^\infty \ldots \int_0^\infty}_{N-|\bar{d}|} \int_0^\infty \frac{(|\bar{d}\,|-1)!}{b^{|\bar{d}|-1}} g(t) \prod_{k:d_k=1} \bar{F}_k(x_k+t)\,dtdx_{k_1}\ldots dx_{k_{N-|\bar{d}|}} \right\}.$$

Applying the formula of integration by parts to the last two sums and canceling the summands, we get:

$$\int_{\tilde{D}_+} P(e,\tilde{D}_-)\rho(de) =$$

$$= \rho_0 \sum_{\bar{d}\in D_+} \left[\sum_{\substack{j:d_j=1,\\ \bar{d}^{(j)}\in D_-}} |\bar{d}|!E^{|\bar{d}|}\beta \prod_{\substack{k:d_k=1,\\ k\neq i}} E\alpha_k + \sum_{\substack{j:d_j=0,\\ \bar{d}^{(j)}\in D_-}} (|\bar{d}|-1)!E^{|\bar{d}|-1}\beta \prod_{k:d_k=1} E\alpha_k \right].$$

Consequently, the QS average stationary sojourn time in D_+ states is determined by the formula:

$$T_{D_+} = \frac{\sum_{\bar{d}\in D_+} |\bar{d}|!E^{|\bar{d}|}\beta \prod_{k:d_k=1} E\alpha_k}{\sum_{\bar{d}\in D_+} \left[\sum_{\substack{j:d_j=1,\\ \bar{d}^{(j)}\in D_-}} |\bar{d}|!E^{|\bar{d}|}\beta \prod_{\substack{k:d_k=1,\\ k\neq j}} E\alpha_k + \sum_{\substack{j:d_j=0,\\ \bar{d}^{(j)}\in D_-}} (|\bar{d}|-1)!E^{|\bar{d}|-1}\beta \prod_{k:d_k=1} E\alpha_k \right]}. \quad (6.22)$$

Specifically, due to (6.22), average stationary time during which N servers are busy is as follows:

$$T_N = \frac{\prod_{k=1}^{N} E\alpha_i}{\sum_{i=1}^{N} \prod_{\substack{k=1,\\ k\neq i}}^{N} E\alpha_k}. \quad (6.23)$$

Consider the example of the formula (6.22) application. In case of three-server QS

$$D = \{(000),(100),(010),(001),(110),(011),(101),(111)\}.$$

Let $D_+ = \{(110),(011),(101)\}$, $D_- = \{(000),(100),(010),(001),(111)\}$, that is, two servers are busy. Then, with the help of (6.22) we get

$$T_{D_+} = E\beta\left(E\alpha_1 E\alpha_2 + E\alpha_1 E\alpha_3 + E\alpha_2 E\alpha_3\right)/\big(2E\beta(E\alpha_1 + E\alpha_2 + E\alpha_3) +$$
$$+ E\alpha_1 E\alpha_2 + E\alpha_1 E\alpha_3 + E\alpha_2 E\alpha_3\big).$$

One should note we can build the semi-Markov model of the $GI/\vec{G}/2/0$ QS by using «residual time». Then the system of integral equations for the EMC stationary distribution determination resembles the system (3.5). And, the solution can be obtained by the method described in Appendix C.

In this case, QS stationary characteristics become more complicated. One can get general idea about them in Section 3.1.

6.2 THE SYSTEM WITH CUMULATIVE RESERVE OF TIME

6.2.1 System Description

In the present section, semi-Markov model of system with cumulative reserve of time is studied [5].

The system *S* includes the object, represented by one structural element, and cumulative reserve of time (nonrandom), which equals $\tau > 0$. The object operating time to failure is RV α with DF $F(t)$ and DD $f(t)$; restoration time is RV β with DF $G(t)$ and DD $g(t)$. System failure occurs when total operating time of the object reaches τ (the system runs out of time reserve) and continues up to the moment of object restoration. Herewith, we consider that by the moment of the object restoration the cumulative reserve of time makes up τ. RVs α, β are assumed to be independent and to have finite expectations and variances.

6.2.2 Semi-Markov Model Building

To describe the system operation we shall use MRP $\{\xi_n, \theta_n; n \geq 0\}$ and the corresponding SMP $\xi(t)$ with the states:

$1x$ – object operating capacity has been restored, the value of time reserve remaining is x, $0 < x \leq \tau$;

$0x$ – the object has passed into down-state, reserve of time remaining is x, $0 < x \leq \tau$;

ωx – system failure occurs, time x, $x > 0$ remains till the complete object restoration.

Under $t = 0$, the system is assumed to be in the state 1τ.

System time diagram is given in Figure 6.1.

Semi-Markov kernel $Q(t, x, B)$ of MRP in a differential form can be written as follows:

$$Q(t,1x,0x) = F(t); \qquad Q(t,0x,1dx_1) = g(x - x_1)1_{x-x_1}(t)dx_1, \qquad 0 < x_1 < x;$$

$$Q(t,0x,\omega dx_1) = g(x + x_1)dx_1 1_x(t).$$

6.2.3 System Characteristics Determination

Let us determine the probability of system S nonfailure operation. Let ζ_{1x}, ζ_{0x} be SMP $\xi(t)$ sojourn times in the subset of up-states $E_+ = \{1x,\ 0x,\ 0 < x \leq \tau\}$ with

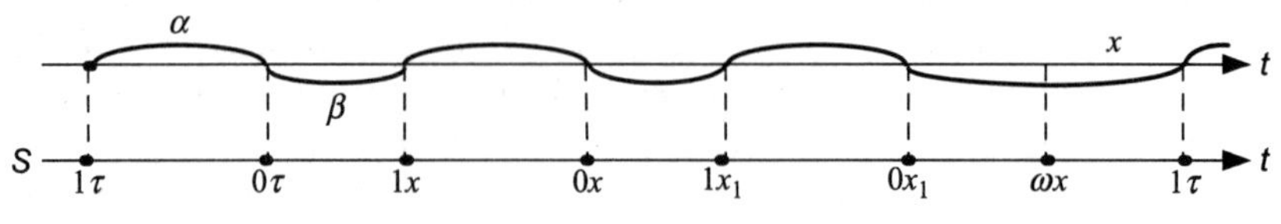

FIGURE 6.1 System operation time diagram

corresponding initial times $1x$, $0x$, and $\phi_1(x,t)$, $\phi_0(x,t)$ are their DFs. The system of Markov renewal equations (1.21) for the functions $\bar{\varphi}_1(x,t)=1-\varphi_1(x,t)$, $\bar{\varphi}_0(x,t)=1-\varphi_0(x,t)$, with regard to $\bar{\varphi}_1(x,t)=\bar{\varphi}_0(x,t)=1$, $0\le t\le x$, can be given as follows:

$$\begin{cases} \bar{\phi}_1(x,t)=\int\limits_x^t f(t-s)\bar{\phi}_0(x,s)ds+\bar{F}(t-x), \quad t\ge x, \\ \bar{\phi}_0(x,t)=\int\limits_0^x g(x-s)\bar{\phi}_1(s,t-x+s)ds, \quad t\ge x. \end{cases} \tag{6.24}$$

Let us introduce the following operators:

$$\left[A_f\varphi\right](x,t)=\int\limits_x^t f(t-s)\varphi(x,s)ds,$$

$$\left[B_g\varphi\right](x,t)=\int\limits_0^x g(x-s)\varphi(s,t-x+s)ds.$$

Then the system (6.24) takes the form:

$$\begin{cases} \bar{\varphi}_1(x,t)=\left[A_f\bar{\varphi}_0\right](x,t)+\bar{F}(t-x), \ t\ge x, \\ \bar{\varphi}_0(x,t)=\left[B_g\bar{\varphi}_1\right](x,t), \ t\ge x, \end{cases} \tag{6.25}$$

consequently,

$$\bar{\varphi}_1(x,t)=\left[\Gamma\bar{\varphi}_1\right](x,t)+\bar{F}(t-x), \ \Gamma=A_fB_g. \tag{6.26}$$

Let us solve the equation (6.26) in Banach space $L(D)$, $D=\{(x,t)\in R^2 \mid 0\le x\le\tau, \ t\ge x\}$ with the norm

$$\|\varphi(x,t)\|=\iint\limits_D |\varphi(x,t)|\,dx\,dt.$$

We can show the norm of some n_0^{th} degree of operator Γ in this space is less than unity. As the operators A_f and B_g are commutative,

$$\left[\Gamma^n\varphi\right](x,t)=\left[(A_f^nB_g^n)\varphi\right](x,t)=$$

$$=\int\limits_x^t f^{*(n)}(t-s)\,ds\int\limits_0^x g^{*(n)}(x-y)\varphi(y,s-x+y)\,dy,$$

$$\left\|\left[\Gamma^n\varphi\right](x,t)\right\|\le\int\limits_0^\tau dx\int\limits_x^\infty dt\int\limits_x^t f^{*(n)}(t-s)\,ds\int\limits_0^x g^{*(n)}(x-y)\left|\varphi(y,s-x+y)\right|\,dy=$$

$$=\int_0^{\tau}dx\int_x^{\infty}ds\int_0^{x}g^{*(n)}(x-y)\left|\varphi(y,s+y-x)\right|dy\int_0^{\infty}f^{*(n)}(t)\,dt=$$

$$=\int_0^{\tau}dx\int_x^{\infty}\left|\varphi(y,s-x+y)\right|ds\int_0^{x}g^{*(n)}(x-y)\,dy=$$

$$=\int_0^{\tau}g^{*(n)}(y)\,dy\int_0^{\tau-y}dx\int_x^{\infty}\left|\varphi(x,t)\right|dt\le G^{*(n)}(\tau)\left\|\varphi(x,t)\right\|.$$

Since the distribution of restoration time is nonarithmetic, there exists n_0, so that $G^{*(n_0)}(\tau)<1$.

Then the solution of the equation (6.26) is determined by the formula:

$$\bar{\varphi}_1(x,t)=(\mathrm{I}-\Gamma)^{-1}\left[\bar{F}(t-x)\right]=\bar{F}(t-x)+\sum_{n=1}^{\infty}\Gamma^n\left[\bar{F}(t-x)\right].$$

Consequently, if under $t=0$ the system was in the state 1τ, the probability of nonfailure system operation is determined by means of the expression:

$$\bar{\phi}_1(x,t)=\begin{cases}1,\ 0\le t<\tau,\\ \bar{F}(t-x)+\sum\limits_{n=1}^{\infty}G^{*(n)}(\tau)\left(\bar{F}^{*(n)}(t-x)-\bar{F}^{*(n+1)}(t-\tau)\right),\ t\ge\tau.\end{cases}\tag{6.27}$$

Let us obtain the average value $E\zeta_{1x}$ and the variance $D\zeta_{1x}$ of the system S time to failure.

$$E\zeta_{1x}=\int_0^{\infty}\bar{\varphi}_1(x,t)\,dt=x+\int_x^{\infty}\bar{F}(t-x)\,dt+$$

$$+\sum_{n=1}^{\infty}G^{*(n)}(x)\int_x^{\infty}\left(F^{*(n)}(t-x)-F^{*(n+1)}(t-x)\right)dt=$$

$$=x+E\alpha+\sum_{n=1}^{\infty}G^{*(n)}(x)\int_x^{\infty}dt\int_x^{t}f^{*(n)}(t-s)\bar{F}(s-x)\,ds=$$

$$=x+E\alpha+\sum_{n=1}^{\infty}G^{*(n)}(x)\int_x^{\infty}\bar{F}(s-x)\,ds\int_s^{\infty}f^{*(n)}(t-s)\,dt=x+E\alpha H(x),$$

where $H(x)=\sum\limits_{n=0}^{\infty}G^{*(n)}(x)$ is the renewal function.

$$D\zeta_{1x}=2\int_0^{\infty}t\bar{\varphi}_1(x,t)dt-E^2\zeta_{1x},$$

$$\int_0^\infty t\bar{\varphi}_1(x,t)\,dt = \int_0^x t\,dt + \int_x^\infty t\bar{F}(t-x)\,dt + \sum_{n=1}^\infty G^{*(n)}(x)\int_x^\infty t\,dt\int_x^t \bar{F}(s-x)f^{*(n)}(t-s)\,ds =$$

$$= \frac{x^2}{2} + xE\alpha + \frac{1}{2}D\alpha + \frac{1}{2}E^2\alpha + \sum_{n=1}^\infty G^{*(n)}(x)\int_x^\infty \bar{F}(s-x)\,ds\int_0^\infty (y+s)f^{*(n)}(y)\,dy =$$

$$= \frac{x^2}{2} + xE\alpha + \frac{1}{2}D\alpha + \frac{1}{2}E^2\alpha + \sum_{n=1}^\infty G^{*(n)}(x)\left(xE\alpha + \frac{1}{2}D\alpha + \frac{1}{2}E^2\alpha\right) +$$

$$+E^2\alpha\sum_{n=1}^\infty nG^{*(n)}(x) = \frac{x^2}{2} + \left(xE\alpha + \frac{1}{2}D\alpha + \frac{1}{2}E^2\alpha\right)H(x) + E^2\alpha\sum_{n=1}^\infty nG^{*(n)}(x).$$

Consequently [8],

$$D\zeta_{1x} = D\alpha H(x) + E^2\alpha\left(\sum_{n=0}^\infty (2n+1)G^{*(n)}(x) - H^2(x)\right) =$$

$$= D\alpha H(x) + E^2\alpha D(N_x) = D\alpha H(x) + E^2\alpha\big(2(H*H)(x) - H(x) - H^2(x)\big),$$

where N_x is the number of jumps of the renewal process generated by DF $G(t)$ on the interval $[0,x]$. In such a way,

$$E\zeta_{1\tau} = \tau + E\alpha H(\tau) = \tau + E\alpha\, E(N_\tau), \tag{6.28}$$

$$D\zeta_{1\tau} = D\alpha\, H(\tau) + E^2\alpha\big(2(H*H)(\tau) - H(\tau) - H^2(\tau)\big) = \tag{6.29}$$

$$= D\alpha\, E(N_\tau) + E^2\alpha\, D(N_\tau).$$

One can find renewal functions $H(x)$, contained in formulas (6.28), (6.29), for different distribution laws, which are in use, in [3, 7].

By applying asymptotic decompositions of the functions $E(N_x)$ and $D(N_x)$ one can obtain approximate equalities for $E\zeta_{1\tau}$, $D\zeta_{1\tau}$ under big values of τ. Let us indicate $\mu = E\beta, \mu_3 = E\beta^3, \sigma^2 = D\beta$. According to [3] we have

$$E(N_\tau) = \frac{\tau}{\mu} + \frac{\sigma^2}{2\mu^2} + \frac{1}{2} + o(1), \tau \to \infty,$$

$$D(N_\tau) = \frac{\sigma^2}{\mu^3}\tau + \left(\frac{1}{12} + \frac{5\sigma^4}{4\mu^4} - \frac{2\mu_3}{3\mu^3}\right) + o(1),\ \tau \to \infty.$$

Consequently,

$$E\zeta_{1\tau} = \left(1 + \frac{E\alpha}{\mu}\right)\tau + E\alpha\left(\frac{\sigma^2}{2\mu^2} + \frac{1}{2}\right) + o(1), \tau \to \infty, \tag{6.30}$$

$$D\zeta_{1\tau} = \left(\frac{D\alpha}{\mu} + \frac{E^2\alpha\sigma^2}{\mu^3}\right)\tau + D\alpha\left(\frac{\sigma^2}{2\mu^2} + \frac{1}{2}\right) + \tag{6.31}$$

$$+E^2\alpha\left(\frac{1}{12} + \frac{5\sigma^4}{4\mu^4} - \frac{2\mu_3}{3\mu^3}\right) + o(1), \quad \tau \to \infty.$$

Example. Let us take the following system with cumulative reserve of time as an example. Its operating time to failure has gamma-distribution with the parameters $\alpha = 30, \beta = 900$; the average operation time to failure of the object is 30 h. Object restoration time has Erlangian distribution of the fourth order with the parameter $\lambda = 6$; the average restoration time of the object equals 0.667 h. In Table 6.1, the values of average operating time to failure, calculated by the formulas (6.28), (6.30) are given.

Now we are to determine the system stationary characteristics. Let us define the EMC $\{\xi_n; n \geq 0\}$ transition probabilities.

$$P_{1x}^{0x} = 1, 0 < x < \tau; \quad P_{1\tau}^{0\tau} = 1; \quad p_{0x}^{1x_1} = g(x - x_1), \quad 0 < x_1 < x < \tau;$$
$$p_{0x}^{\omega x_1} = g\,(x + x_1), \quad x_1 > 0; \quad P_{\omega x}^{1\tau} = 1; \quad p_{0\tau}^{1x} = g(\tau - x), \quad 0 < x < \tau;$$
$$p_{0\tau}^{\omega x} = g\,(\tau + x).$$

TABLE 6.1 Dependence of Average Operating Time to Failure of Cumulative System on the Accumulator Capacity

τ, h	$E\zeta_{1\tau}$, h (6.28)	$E\zeta_{1\tau}$, h (6.30)	τ, h	$E\zeta_{1\tau}$, h (6.28)	$E\zeta_{1\tau}$, h (6.30)
0	30	18.75	1.3	78.55	78.55
0.1	30.20	23.35	1.4	83.15	83.15
0.2	31.21	27.95	1.5	87.75	87.75
0.3	33.58	32.55	1.6	92.35	92.35
0.4	37.14	37.15	1.7	96.95	96.95
0.5	41.44	41.75	1,8	101.55	101.55
0.6	46.08	46.35	1.9	106.15	106.15
0.7	50.79	50.95	2.0	110.75	110.75
0.8	55.49	55.55	2.1	115.35	115.30
0.9	60.15	60.15	2.2	119.95	119.95
1.0	64.76	64.75	2.3	124.55	124.55
1.1	69.36	69.35	2.4	129.55	129.55
1.2	73.96	73.95	2.5	133.75	133.75

The system of equations for the EMC stationary distribution looks like:

$$\begin{cases} \rho(1x) = \int\limits_x^\tau g(s-x)\rho(0s)\,ds + \rho(0\tau)g(\tau - x), \ 0 < x < \tau, \\ \rho(0x) = \rho(1x), \quad 0 < x < \tau, \\ \rho(0\tau) = \rho(1\tau), \\ \rho(1\tau) = \int\limits_0^\infty \rho(\omega x)\,dx, \\ \rho(\omega x) = \int\limits_0^\tau g(s+x)\rho(0s)\,ds + \rho(0\tau)g(\tau + x), \ x > 0, \\ \rho(1\tau) + \rho(0\tau) + \int\limits_0^\tau \left(\rho(1x) + \rho(0x)\right)dx + \int\limits_0^\infty \rho(\omega x)\,dx = 1. \end{cases} \tag{6.32}$$

Excluding $\rho\,(0x)$ and $\rho\,(0\tau)$ from the first equation of the system (6.32), we get the following equation

$$\rho\,(1x) = \int\limits_x^\tau g(s-x)\rho(1s)\,ds + \rho\,(1\tau)g(\tau - x).$$

Its solution is

$$\rho(1x) = \rho(1\tau)h(\tau - x), \tag{6.33}$$

where $h(x) = \sum\limits_{n=1}^{\infty} g^{*(n)}(x)$ is the density of renewal function $H(x)$.

The rest of stationary densities are determined by the formulas:

$$\rho(0x) = \rho(1x), \quad \rho(\omega x) = \rho(1\tau)\int\limits_0^\tau g(s+x)h(\tau - s)\,ds + \rho(1\tau)g(\tau + x). \tag{6.34}$$

The constant $\rho\,(1\tau)$ can be obtained from the normalization requirement. By transforming the left-hand side of the last equation of the system (6.32), we get

$$2\rho(1\tau)H(\tau) + \rho(1\tau)\int\limits_0^\tau h(\tau - s)\,\bar{G}(s)\,ds + \rho\,(1\tau)\bar{G}(\tau) = \rho(1\tau)\left(1 + 2H(\tau)\right).$$

Consequently, $\rho(1\tau) = \left(1 + 2H(\tau)\right)^{-1}$.

Apart from the above subset of up-states E_+, let us consider the subset of system down-states $E_- = \{\omega x, x > 0\}$.

Let us define the averages of system sojourn times defined as follows:

$$\theta_{1x} = \theta_{1\tau} = \alpha, \theta_{0x} = \beta \wedge x, \theta_{0\tau} = \beta \wedge \tau, \theta_{\omega x} = x,$$

consequently,

$$m(1x) = m(1\tau) = E\alpha, m(0x) = \int\limits_0^x \bar{G}(t)\,dt, m(0\tau) = \int\limits_0^\tau \bar{G}(t)\,dt, m(\omega x) = x. \tag{6.35}$$

Let us obtain the system stationary availability factor K_a. To do it we apply formulas (6.33)–(6.35) and calculate the integrals

$$\int_{E_+} m(e)\rho(de) = \rho(1\tau)\left(E\alpha H(\tau) + \int_0^{\tau} h(\tau - x)\,dx \int_0^{x} \bar{G}(t)\,dt + \int_0^{\tau} \bar{G}(t)\,dt \right) =$$

$$= \rho(1\tau)(E\alpha H(\tau) + \tau),$$

$$\int_{E_-} m(e)\rho(de) = \rho(1\tau)\int_0^{\infty} x\,dx \left(\int_0^{\tau} g\,(s + x)h(\tau - s)\,ds + g(\tau + x) \right) =$$

$$= \rho(1\tau)\left(\int_0^{\infty} dt \int_t^{\infty} dx \int_0^{\tau} g(s + x)h(\tau - s)\,ds + \int_0^{\infty} dt \int_t^{\infty} g(\tau + x)\,dx) \right) =$$

$$= \rho(1\tau)\left(\int_0^{\tau} h(\tau - s)\,ds \int_s^{\infty} \bar{G}(t)\,dt + \int_{\tau}^{\infty} \bar{G}(t)\,dt \right) = \rho(1\tau)\left(E\beta H(\tau) - \right.$$

$$\left. - \left(\int_0^{\tau} h(\tau - s)\,ds \int_0^{s} \bar{G}(t)\,dt + \int_0^{\tau} \bar{G}(t)\,dt \right) \right) = \rho(1\tau)\big(E\beta H(\tau) - \tau\big),$$

$$\int_E m(e)\rho(de) = \rho(1\tau)\big(E\alpha + E\beta\big)H(\tau).$$

Consequently, applying the formula (1.16), we have

$$K_a = \frac{\tau + E\alpha H(\tau)}{(E\alpha + E\beta)H(\tau)}.$$

Let us determine operating time to failure T_+ and average stationary restoration time T_- of the system S.

$$\int_{E_+} \rho(de)P(e, E_-) = \int_0^{\tau} G(s)\rho(0s)\,ds + \rho(1\tau)\bar{G}(\tau) =$$

$$= \rho(1\tau)\left(\int_0^{\tau} h(\tau - s)\bar{G}(s)\,ds + \bar{G}(\tau) \right) = \rho(1\tau),$$

$$\int_{E_-} \rho(de)P(e, E_+) = \int_0^{\infty} \rho(\omega x)\,dx = \rho(1\tau).$$

In such a way, on the basis of (1.17), (1.18) we have:

$$T_+ = \tau + E\alpha H(\tau), \quad T_- = E\beta H(\tau) - \tau.$$

Based on the results obtained let us determine the value of time reserve ensuring necessary system reliability.

6.3 TWO-PHASE SYSTEM WITH A INTERMEDIATE BUFFER

6.3.1 System Description

In the present section, two-phase system with a intermediate buffer is considered [5, 10]. Its block scheme is given in Figure 6.2.

Let us describe the system operation. The input of the first device A_1 of productivity c_1 is $N_1(\Delta t)$ production units (as many as the device can process within this time period, that is, $N_1(\Delta t) = c_1\Delta t$) per time unit Δt. Having being processed by the first device, the production goes to the second device A_2, $N_2(\Delta t) = c_2\Delta t$. Operating time to failure (restoration time) of the device A_i, $i = 1, 2$, is RV $\alpha_i^{(1)}$ $(\alpha_i^{(0)})$ with DF $F_i^{(1)}(t)$ $(F_i^{(0)}(t))$ and DD $f_i^{(1)}(t)$ $(f_i^{(0)}(t))$. RVs $\alpha_i^{(1)}, \alpha_i^{(0)}$ are assumed to be independent and to have finite expectations. The capacity of absolutely reliable buffer stock H is expressed in terms of units of time, which is necessary for the device A_2 to empty a full buffer. The buffer capacity equals $h \ge 0$. If device A_2 fails and the buffer is full, A_1 switches off. If device A_1 fails and the buffer is empty, the device A_2 switches off. The quantity of production is supposed to be a continuous value (the discontinuity of product units should be taken into account under a small buffer capacity). Restoration of devices A_1, A_2 is supposed to be infinite. The whole system is considered to be operative if there is production at the output of device A_2.

Thus, at the output of A_1 there is independent time reserve, which can be completed either by means of stock of A_1-device speed or by standby time of A_2. Let us call the above system real.

Suppose the devices A_1 and A_2 have nonincreasing productivity $(c_1 \ge c_2)$.

6.3.2 Semi-Markov Model Building

To describe the operation of the real system, let us introduce the following space of semi-Markov states:

$$E = \left\{ i\bar{d}\bar{x}z : \; i = 1, 2, \;\; \bar{d} = (d_1, d_2), \;\; \bar{x} = (x_1, x_2) \right\},$$

where i is the number of device A_i, which was last to fail or to be restored; the component d_k of vector $\bar{d}$ indicates the state of A_k: up-state $(d_k = 1)$, restoration $(d_k = 0)$, down-state $(d_k = 2)$. The component x_k of vector $\bar{x}$ is time, which passed since the moment of the last restoration or failure of device A_k, $x_i = 0$. The value z defines time reserve of buffer H, $0 \le z \le h$.

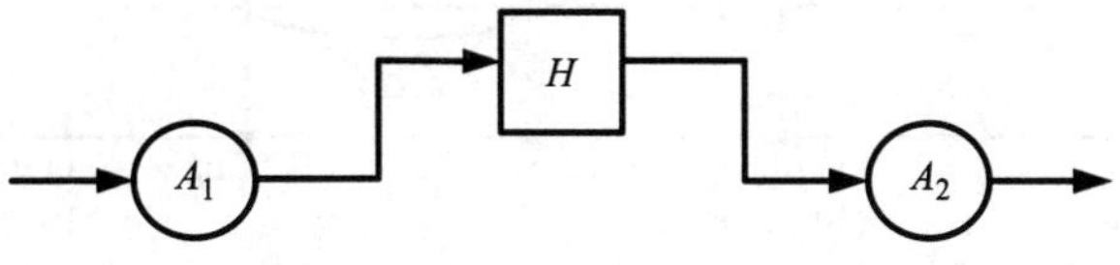

FIGURE 6.2 Block scheme of two-phase system with a intermediate buffer

It is possible to completely build the SMP describing real system operation. In our case only SMP characteristics, which are used in the method applied, are defined.

6.3.3 System Stationary Characteristics Approximation

To approximate real system stationary characteristics we apply the method based on formulas (1.25). Let us choose support system $S^{(0)}$. Suppose the device A_1 has a quick restoration, that is, its restoration time $\alpha_1^{(0)}$ depends on a small positive parameter ε, so that:

$$\lim_{\varepsilon\to 0} E\alpha_1^{(0,\varepsilon)} = 0, \tag{6.36}$$

However, an output device A_2 operating time to failure and restoration time are fixed. It results in the following support system $S^{(0)}$. Its device A_1 is restored immediately, and the buffer H is completely full. Time diagram of support system operation is given in Figure 6.3.

Let us define EMC transition probabilities of support system:

$$p_{1110x_2h}^{1010y_2h} = \frac{f_1^{(1)}(y_2 - x_2)\bar{F}_2^{(1)}(y_2)}{\bar{F}_2^{(1)}(x_2)}, y_2 > x_2, \; p_{1110x_2h}^{221y_10h} = \frac{f_2^{(1)}(y_1 + x_2)\bar{F}_1^{(1)}(y_1)}{\bar{F}_2^{(1)}(x_2)}, y_1 > 0;$$

$$p_{211x_10h}^{220y_10h} = \frac{f_2^{(1)}(y_1 - x_1)\bar{F}_1^{(1)}(y_1)}{\bar{F}_1^{(1)}(x_1)}, y_1 > x_1, \quad p_{211x_10h}^{1010y_2h} = \frac{f_1^{(1)}(y_2 + x_1)\bar{F}_2^{(1)}(y_2)}{\bar{F}_1^{(1)}(x_1)}, y_1 > 0;$$

$$\mathrm{P}_{1010x_2h}^{1110x_2h} = \mathrm{P}_{220x_10h}^{211x_10h} = 1.$$

In [14] the density of the EMC stationary distribution is shown to be as follows:

$$\begin{aligned} \rho(1110x_2h) &= \rho(1010x_2h) = \rho_0\bar{F}_2^{(1)}(x_2), \\ \rho(211x_10h) &= \rho(220x_10h) = \rho_0\bar{F}_1^{(1)}(x_1), \end{aligned} \tag{6.37}$$

where constant ρ_0 is found explicitly from the normalization requirement.

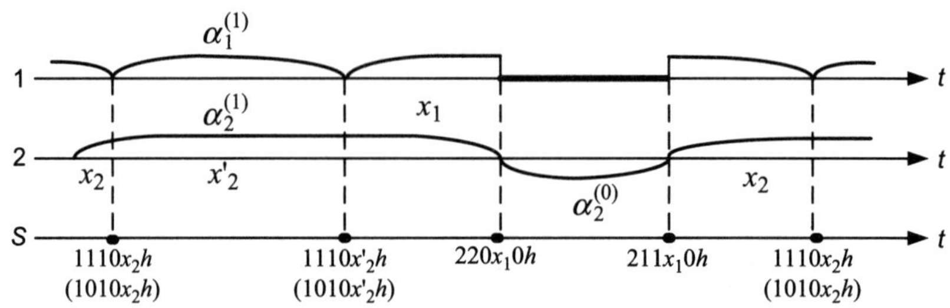

FIGURE 6.3 Time diagram of support system operation

So, the class of ergodic states $E^{(0)}$ of support system $S^{(0)}$ includes the following states:

$$E^{(0)} = \{1110x_2h,\ \ 1010x_2h,\ \ 211x_10h,\ \ 220x_10h\}.$$

For the real system with the above-mentioned criterion of the system failure, the subsets of up (E_+) and down (E_-) states are:

$$E_+ = \{1110x_2z,\ \ 1010x_2z,\ \ 211x_10z,\ \ 201x_10z\},$$

$$E_- = \{1100x_2z,\ \ 1000x_2z,\ \ 1020x_20,\ \ 120x_1x_2h,\ \ 210x_10z,\ \ 200x_10z,\ \ 202x_1x_20,\ \ 220x_10h\}.$$

Note that such states with discrete codes as 112, 121, 122, 212, 221, 222 are not used in description of the real system.

Let us determine the expressions from the formulas (1.25). For the system considered, $r = 1$, as the real system can transit from up-states (of the ergodic class $E^{(0)}$) into the subset of down-states E_- in one step.

$$\int_{E_+} \mathrm{P}(x,E_-)\rho(dx) = \int_0^\infty \rho(1110x_2h)\mathrm{P}(1110x_2h,E_-)dx_2 +$$

$$+\int_0^\infty \rho(1010x_2h)\mathrm{P}(1010x_2h,E_-)dx_2 + \int_0^\infty \rho(211x_10h)\mathrm{P}(211x_10h,E_-)dx_1 =$$

$$= \rho_0\left(\int_0^\infty \bar{F}_1^{(1)}(y_1)\bar{F}_2^{(1)}(y_1)dy_1 + \bar{F}_1^{(0,\varepsilon)}(h)\int_h^\infty \bar{F}_2^{(1)}(x_2)dx_2 +\right.$$

$$\left.+\int_0^h \bar{F}_1^{(0,\varepsilon)}(y_1)dy_1\int_0^\infty f_2^{(1)}(x_2+y_1)dx_2 + \int_0^\infty \bar{F}_1^{(1)}(y_1)dy_1\int_0^{y_1} f_2^{(1)}(y_1-x_1)dx_1\right) =$$

$$= \rho_0\left(E\alpha_1^{(1)} + \bar{F}_1^{(0,\varepsilon)}(h)\int_h^\infty \bar{F}_2^{(1)}(x_2)dx_2 + \int_0^h \bar{F}_1^{(0,\varepsilon)}(y_1)\bar{F}_2^{(1)}(y_1)dy_1\right).$$

Let us obtain average sojourn times $m(x)$ of the real system in ergodic states:

$$\theta_{1110x_2h} = \alpha_1^{(1)} \wedge [\alpha_2^{(1)} - x_2]^+,\quad m(1110x_2h) = \int_0^\infty \frac{\bar{F}_1^{(1)}(t)\bar{F}_2^{(1)}(x_2+t)}{\bar{F}_2^{(1)}(x_2)}dt,$$

$$\theta_{1010x_2h} = \alpha_1^{(0,\varepsilon)} \wedge [\alpha_2^{(1)} - x_2]^+ \wedge h,\quad m(1010x_2h) = \int_0^h \frac{\bar{F}_1^{(0,\varepsilon)}(t)\bar{F}_2^{(1)}(x_2+t)}{\bar{F}_2^{(1)}(x_2)}dt,$$

$$\theta_{211x_10h} = \alpha_2^{(1)} \wedge [\alpha_1^{(1)} - x_1]^+,\quad m(211x_10h) = \int_0^\infty \frac{\bar{F}_2^{(1)}(t)\bar{F}_1^{(1)}(x_1+t)}{\bar{F}_1^{(1)}(x_1)}dt.$$

Consequently,

$$\int_{E_+} m(x)\rho(dx) = \rho_0 \left(\int_0^\infty dx_2 \int_0^\infty \bar{F}_1^{(1)}(t)\bar{F}_2^{(1)}(x_2+t)\,dt + \right.$$

$$\left. + \int_0^\infty dx_2 \int_0^h \bar{F}_1^{(0,\varepsilon)}(t)\bar{F}_2^{(1)}(x_2+t)\,dt + \int_0^\infty dx_2 \int_0^\infty \bar{F}_2^{(1)}(t)\bar{F}_1^{(1)}(x_1+t)\,dt \right) =$$

$$= \rho_0 \left(E\alpha_1^{(1)} E\alpha_2^{(1)} + \int_0^h \bar{F}_1^{(0,\varepsilon)}(t)\,dt \int_0^\infty \bar{F}_2^{(1)}(x_2+t)\,dx_2 \right).$$

To prove the latter equality we use the following statement [19]:

Lemma. Let $F_j(t)$, $j=\overline{1,n}$ be DFs of independent nonnegative RVs α_j with finite expectations. Then the following equality is true:

$$\sum_{j=1}^{n} \int_0^\infty \ldots \int_0^\infty \bar{F}_j(t) \left(\prod_{\substack{k=1,\\ k\neq j}}^{n} \bar{F}_k(t+y_k)\,dy_k \right) dt = \prod_{j=1}^{n} E\alpha_j. \tag{6.38}$$

In such a way, average operation time to failure T_+ of the real system we consider can be approximated in the following way:

$$T_+ \approx \left(E\alpha_1^{(1)} E\alpha_2^{(1)} + \int_0^h \bar{F}_1^{(0,\varepsilon)}(t)\,dt \int_0^\infty \bar{F}_2^{(1)}(x_2+t)\,dx_2 \right) \Bigg/$$
$$\Bigg/ \left(E\alpha_1^{(1)} + \bar{F}_1^{(0,\varepsilon)}(h) \int_h^\infty \bar{F}_2^{(1)}(x_2)\,dx_2 + \int_0^h \bar{F}_1^{(0,\varepsilon)}(y_1)\bar{F}_2^{(1)}(y_1)\,dy_1 \right). \tag{6.39}$$

Due to the condition (6.36)

$$\int_0^h \bar{F}_1^{(0,\varepsilon)}(y_1)\bar{F}_2^{(1)}(y_1)\,dy_1 = E(\alpha_1^{(0,\varepsilon)} \wedge \alpha_2^{(1)} \wedge h) \approx E(\alpha_1^{(0,\varepsilon)} \wedge h),$$

$$\int_0^h \bar{F}_1^{(0,\varepsilon)}(t)\,dt \int_t^\infty \bar{F}_2^{(1)}(x_2)\,dx_2 = \int_0^h \bar{F}_2^{(1)}(x_2)\,dx_2 \int_0^{x_2} \bar{F}_1^{(0,\varepsilon)}(t)\,dt +$$

$$+ \int_h^\infty \bar{F}_2^{(1)}(x_2)\,dx_2 \int_0^h \bar{F}_2^{(0,\varepsilon)}(t)\,dt \approx \int_0^h \bar{F}_2^{(1)}(x_2)\,dx_2 \int_0^h \bar{F}_1^{(0,\varepsilon)}(t)\,dt +$$

$$+ \int_h^\infty \bar{F}_2^{(1)}(x_2)\,dx_2 \int_0^h \bar{F}_2^{(0,\varepsilon)}(t)\,dt = E\alpha_2^{(1)} \int_0^h \bar{F}_1^{(0,\varepsilon)}(t)\,dt.$$

That is why apart from the formula (6.39) we can apply the following expression for T_+:

$$T_+ \approx \left(E\alpha_1^{(1)} E\alpha_2^{(1)} + E\alpha_2^{(1)} \int_0^h \bar{F}_1^{(0,\varepsilon)}(t)\,dt \right) \Big/ \left(E\alpha_1^{(1)} + \right.$$
$$\left. + \bar{F}_1^{(0,\varepsilon)}(h) \int_h^\infty \bar{F}_2^{(1)}(t)\,dt + \int_0^h \bar{F}_1^{(0,\varepsilon)}(t)\,dt \right), \tag{6.40}$$

and

$$T_+ \approx \left(E\alpha_1^{(1)} E\alpha_2^{(1)} \right) \Big/ \left(E\alpha_1^{(1)} + \bar{F}_1^{(0,\varepsilon)}(h) \int_h^\infty \bar{F}_2^{(1)}(t)\,dt \right). \tag{6.41}$$

Let us determine average stationary restoration time T_-. The denominator of (1.25) was found while calculating T_+. Let us get the numerator of this formula:

$$\int_E \rho(dx) \int_{E_-} m(y) \mathrm{P}(x,dy) = \rho_0 \left(E\alpha_2^{(0)} \int_0^\infty dx_2 \int_0^\infty f_2^{(1)}(x_2+y_1) \bar{F}_1^{(1)}(y_1)\,dy_1 + \right.$$
$$+ \int_0^\infty \bar{F}_2^{(1)}(x_2+h)\,dx_2 \int_0^\infty \bar{F}_1^{(0,\varepsilon)}(h+t)\,dt +$$
$$+ \int_0^\infty dx_2 \int_0^h f_2^{(1)}(x_2+y_1)\,dy_1 \int_0^\infty \bar{F}_2^{(0)}(t) \bar{F}_1^{(0,\varepsilon)}(y_1+t)\,dt +$$
$$\left. + E\alpha_2^{(0)} \int_0^\infty dx_1 \int_{x_1}^\infty f_2^{(1)}(y_1-x_1) \bar{F}_1^{(1)}(y_1)\,dy_1 \right) = \rho_0 \left(E\alpha_1^{(1)} E\alpha_2^{(0)} + \right.$$
$$\left. + \int_h^\infty \bar{F}_2^{(1)}(x_2)\,dx_2 \int_h^\infty \bar{F}_1^{(0,\varepsilon)}(t)\,dt + \int_0^h \bar{F}_2^{(1)}(y_1)\,dy_1 \int_0^\infty \bar{F}_2^{(0)}(t) \bar{F}_1^{(0,\varepsilon)}(y_1+t)\,dt \right).$$

That is why

$$T_- \approx \left(E\alpha_1^{(1)} E\alpha_2^{(0)} + \int_h^\infty \bar{F}_2^{(1)}(x_2)\,dx_2 \int_h^\infty \bar{F}_1^{(0,\varepsilon)}(t)\,dt + E\alpha_2^{(0)} \int_0^h \bar{F}_1^{(0,\varepsilon)}(y_1)\,dy_1 \right) \Big/$$
$$\Big/ \left(E\alpha_1^{(1)} + \bar{F}_1^{(0,\varepsilon)}(h) \int_h^\infty \bar{F}_2^{(1)}(x_2)\,dx_2 + \int_0^h \bar{F}_1^{(0,\varepsilon)}(y_1)\,dy_1 \right). \tag{6.42}$$

Taking into account that

$$\int_0^h \bar{F}_2^{(1)}(y_1)\,dy_1 \int_0^\infty \bar{F}_2^{(0)}(t) \bar{F}_1^{(0,\varepsilon)}(y_1+t)\,dt = \int_0^\infty \bar{F}_2^{(0)}(t)\,dt \int_0^h \bar{F}_2^{(1)}(y_1) \bar{F}_1^{(0,\varepsilon)}(y_1+t)\,dy_1 \approx$$

$$\approx \int_0^\infty \overline{F}_2^{(0)}(t)dt \int_0^h \overline{F}_2^{(1)}(y_1)\overline{F}_1^{(0,\varepsilon)}(y_1)dy_1 \approx E\alpha_2^{(0)} \int_0^h \overline{F}_1^{(0,\varepsilon)}(y_1)dy_1,$$

we also get

$$T_- \approx \left(E\alpha_1^{(1)}E\alpha_2^{(0)} + \int_h^\infty \overline{F}_2^{(1)}(x_2)dx_2 \int_h^\infty \overline{F}_1^{(0,\varepsilon)}(t)dt + E\alpha_2^{(0)} \int_0^h \overline{F}_1^{(0,\varepsilon)}(y_1)dy_1 \right) \Bigg/$$

$$\Bigg/ \left(E\alpha_1^{(1)} + \overline{F}_1^{(0,\varepsilon)}(h) \int_h^\infty \overline{F}_2^{(1)}(x_2)dx_2 + \int_0^h \overline{F}_1^{(0,\varepsilon)}(y_1)dy_1 \right), \tag{6.43}$$

and

$$T_- \approx \left(E\alpha_1^{(1)}E\alpha_2^{(0)} + \int_h^\infty \overline{F}_2^{(1)}(x)dx \int_h^\infty \overline{F}_1^{(0,\varepsilon)}(t)dt \right) \Bigg/$$

$$\Bigg/ \left(E\alpha_1^{(1)} + \overline{F}_1^{(0,\varepsilon)}(h) \int_h^\infty \overline{F}_2^{(1)}(t)dt \right). \tag{6.44}$$

By applying formulas (6.38)–(6.43), we can get stationary availability factor of the real system:

$$K_a \approx \left(E\alpha_1^{(1)}E\alpha_2^{(1)} + E\alpha_2^{(1)} \int_0^h \overline{F}_1^{(0,\varepsilon)}(t)dt \right) \Bigg/ \Bigg(E\alpha_1^{(1)}E\alpha_2^{(1)} +$$

$$+E\alpha_1^{(1)}E\alpha_2^{(0)} + E\alpha_2^{(1)} \int_0^h \overline{F}_1^{(0,\varepsilon)}(t)dt + E\alpha_2^{(0)} \int_0^h \overline{F}_1^{(0,\varepsilon)}(t)dt + \tag{6.45}$$

$$+\int_h^\infty \overline{F}_2^{(1)}(x)dx \int_h^\infty \overline{F}_1^{(0,\varepsilon)}(t)dt \Bigg),$$

and

$$K_a \approx \left(E\alpha_1^{(1)}E\alpha_2^{(1)} \right) \Bigg/ \Bigg(E\alpha_1^{(1)}E\alpha_2^{(1)} + E\alpha_1^{(1)}E\alpha_2^{(0)} + \tag{6.46}$$

$$+\int_h^\infty \overline{F}_2^{(1)}(x)dx \int_h^\infty \overline{F}_1^{(0,\varepsilon)}(t)dt \Bigg).$$

With the help of stationary availability factor K_a, one can approximate the productivity P of two-phase system with a buffer stock:

$$P = K_a c_2,$$

where c_2 is the productivity of output device A_2.

TABLE 6.2 Reliability Characteristics of Two-phase System With a Buffer

h, (h)	T_+, h (6.40)	T_+, h (6.41)	T_-, h (6.43)	T_-, h (6.44)	K_a (6.45)	K_a (6.46)
0	20.000	20.000	1.700	1.700	0.922	0.922
0.1	20.077	20.052	1.652	1.625	0.924	0.924
0.2	20.209	20.159	1.610	1.609	0.926	0.926
0.3	20.394	20.32	1.571	1.571	0.928	0.928
0.4	20.633	20.535	1.537	1.536	0.931	0.930
0.5	20.926	20.805	1.507	1.506	0.933	0.932
0.6	21.272	21.128	1.482	1.480	0.935	0.935
0.7	21.670	21.505	1.460	1.458	0.937	0.937
0.8	22.121	21.935	1.442	1.44	0.939	0.938
0.9	22.621	22.417	1.428	1.425	0.941	0.940
1.0	23.170	22.949	1.417	1.417	0.942	0.942
1.1	23.766	23.53	1.410	1.406	0.944	0.944
1.2	24.406	24.157	1.406	1.401	0.946	0.945
1.3	25.087	24.826	1.405	1.400	0.947	0.947
1.4	25.804	25.535	1.407	1.401	0.948	0.948
1.5	26.552	26.278	1.412	1.406	0.950	0.949

Example. Let operating times to failure of devices A_1, A_2 of the same productivity have Erlangian distribution of the fourth order with parameter $\lambda = 0.1$, that is, their average operating time to failure equals 40 h. Restoration times of devices have Rayleigh distribution with parameter $\sigma = 1.356$; the average restoration time of the devices is 1.7 h.

In Table 6.2, the values of T_+, T_-, K_a, calculated by means of formulas (6.40), (6.41), (6.43)–(6.46), are given.

The method considered can be applied to modeling multiphase systems with buffers of different structures.

6.4 THE MODEL OF TECHNOLOGICAL CELL WITH NONDEPRECIATORY FAILURES

6.4.1 System Description

The present section deals with the semi-Markov model of technological cell (TC) with nondepreciatory failures.

Let us describe TC operation. Service time of production unit by TC is RV α_1 with DF $F_1(t) = P\{\alpha_1 \le t\}$ and DD $f_1(t)$. TC operation time to failure is RV α_2 with DF $F_2(t) = P\{\alpha_2 \le t\}$ and DD $f_2(t)$, TC restoration time is RV β_2 with DF $G_2(t) = P\{\beta_2 \le t\}$ and DD $g_2(t)$. When TC fails, the service of production unit is interrupted. After TC restoration, the service continues with regard to the last service time. RVs α_1, α_2, β_2 are assumed to be independent and to have finite expectations and variances.

We have to determine DF $F_\theta(t)$, expectation and variance of RV θ-time of production unit service cycle, and TC productivity. Herewith, TC failures are taken into account.

6.4.2 TC Semi-Markov Model Building

To describe TC operation let us introduce the following set E of semi-Markov system states:

$$E = \{11x, 10x, 21x, 20x\}.$$

The meaning of state codes is the following:

$11x$ – TC is in up-state; it has begun to serve a production unit; time $x > 0$ remains till TC failure;
$10x$ – immediate state corresponding the moment of the end of production unit service; time $x > 0$ remains till TC failure;
$21x$ – TC has been restored; interrupted service of production unit continues; time $x > 0$ remains till the end of the service;
$20x$ – TC has failed; production unit service is interrupted; time $x > 0$ remains till the end of the service.

System operation time diagram and system transition graph are given in Figures 6.4 and 6.5 correspondingly.

Let us determine semi-Markov kernel $Q(t,x,B)$ $\{\xi_n, \theta_n; n \ge 0\}$ in a differential form

$$Q(t, 11x, 10dy) = f_1(x-y)1_{x-y}(t)dy, \quad 0 < y < x,$$

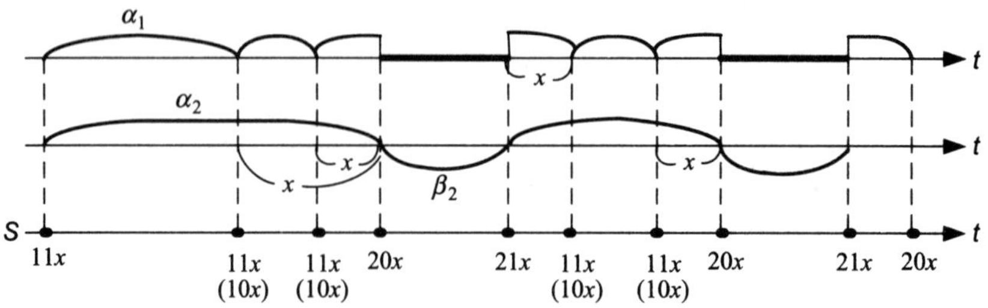

FIGURE 6.4 System operation time diagram

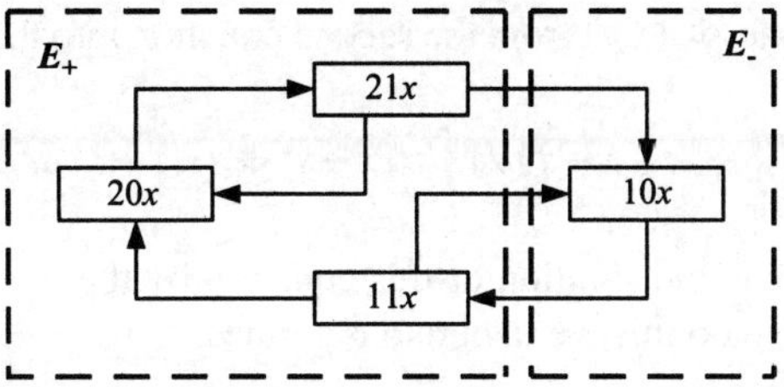

FIGURE 6.5 System transition graph

where $1_{x-y}(t)$ is a unitary distribution at the point $x-y$,

$$Q(t, 11x, 20\, dy) = f_1(x+y)1_x(t)\, dy, \quad y>0,$$

$$Q(t, 21x, 20\, dy) = f_2(x-y)1_{x-y}(t)\, dy, \quad 0<y<x,$$

$$Q(t, 21x, 10\, dy) = f_2(x+y)1_x(t)\, dy, \quad y>0,$$

$$Q(t, 10x, 11x) = 1(t), \quad Q(t, 20x, 21x) = G_2(t).$$

6.4.3 TC Characteristics Determination

Let us introduce the set E of system states in the following way:

$$E = E_+ \cup E_-\ E_+ = \{11x, 21x, 20x\}, E_- = \{10x\}.$$

SMP $\xi(t)$ sojourn time in the subset E_+, under the condition of initial state ($t=0$) $10x$, equals service time of the production unit by TC.

Now we determine the distribution of SMP $\xi(t)$ sojourn times in the subset E_+.

Let $\tau_{10x}, \tau_{20x}, \tau_{21x}$ be sojourn times of $\xi(t)$ in E_+ with the initial state $10x$, $20x$, $21x$ correspondingly; $\Phi_1(x,t), \Phi_2(x,t), \Phi_3(x,t)$ are their DFs.

Let us construct the system of Markov renewal equations (1.21) for the functions $\bar{\Phi}_i(x,t) = 1-\Phi_i(x,t),\ i=\overline{1,3}$:

$$\begin{cases} \bar{\Phi}_2(x,t) = \bar{\Phi}_3(x,t) = 1, \quad \bar{\Phi}_1(x,t) = \bar{F}_1(t) \ \text{if} \ 0 \le t < x, \\ \bar{\Phi}_1(x,t) = \int\limits_0^{t-x} f_1(x+y)\, \bar{\Phi}_3(y, t-x)\, dy + \bar{F}_1(t), \ t \ge x, \\ \bar{\Phi}_2(x,t) = \int\limits_0^{x} f_2(x-s)\, \bar{\Phi}_3(s, t-x+s)\, ds, \ t \ge x, \\ \bar{\Phi}_3(x,t) = \int\limits_0^{t-x} g_2(t-x-\tau)\, \bar{\Phi}_2(x, x+\tau)\, d\tau + \bar{G}_2(t-x), \ t \ge x. \end{cases} \tag{6.47}$$

Let us substitute $\bar{\Phi}_2(x,t)$ from the second equation into the third one. We get

$$\bar{\Phi}_3(x,t)=\int_0^{t-x} g_2(t-x-\tau)\,d\tau\int_0^x f_2(x-s)\,\bar{\Phi}_3(s,\tau+s)\,ds+\bar{G}_2(t-x).$$

Let us determine the solution of this equation by the method of successive approximations. To do this we integrate it n times.

$$\bar{\Phi}_3(x,t)=\int_0^{t-x} g_2^{*(n)}(t-x-\tau)\,d\tau\int_0^x f_2^{*(n)}(x-s)\,\bar{\Phi}_3(s,\tau+s)\,ds+\bar{G}_2(t-x)+$$

$$+\sum_{k=1}^{n-1} F_2^{*(k)}(x)\left(G_2^{*(k)}(t-x)-G_2^{*(k+1)}(t-x)\right),$$

where $g_2^{*(n)}(x), f_2^{*(n)}(x)$ are n-fold convolutions of DDs, and $F_2^{*(k)}(x), G_2^{*(k)}(x)$ are k-fold convolutions of DFs.

Let us pass to the limit under $n\to\infty$ in both parts of the equality. The limit of integral on the right-hand side equals 0 under all x and t as

$$\int_0^{t-x} g_2^{*(n)}(t-x-\tau)\,d\tau\int_0^x f_2^{*(n)}(x-s)\,\bar{\Phi}_3(s,\tau+s)\,ds\le F_2^{*(n)}(x)\,G_2^{*(n)}(t-x)\le F_2^{*(n)}(x),$$

$F_2^{*(n)}(x)\to 0$ under $n\to\infty$.

In such a way, the solution of the system (6.47) is:

$$\bar{\Phi}_3(x,t)=\bar{G}_2(t-x)+\sum_{n=1}^{\infty} F_2^{*(n)}(x)\left(G_2^{*(n)}(t-x)-G_2^{*(n+1)}(t-x)\right),\ t\ge x,$$

$$\bar{\Phi}_2(x,t)=F_2(x)\overline{G_2}(t-x)+\sum_{n=1}^{\infty} F_2^{*(n+1)}(x)\left(G_2^{*(n)}(t-x)-G_2^{*(n+1)}(t-x)\right),\ t\ge x,$$

(6.48)

$$\bar{\Phi}_1(x,t)=\int_0^{t-x} f_1(x+y)\,\bar{G}_2(t-x-y)\,dy+\bar{F}_1(t)+$$

$$+\sum_{n=1}^{\infty}\int_0^{t-x} f_1(x+y)F_2^{*(n)}(y)\left(G_2^{*(n)}(t-x-y)-G_2^{*(n+1)}(t-x-y)\right)dy,\ t\ge x.$$

Formula (6.48) determines DF $\bar{\Phi}_1(x,t)$ of cycle time of production unit service by TC. This DF depends on the initial state $11x$ with a continuous component x. To determine DF $F_\theta(t)$ of RV θ - cycle time of production unit service by TC, regardless of the initial state, we are averaging by the formula:

$$\bar{F}_\theta(t)=\frac{\int_{E_-}\rho(dx)\int_{E_+}\bar{F}_y(t)P(x,dy)}{\int_{E_-}P(x,E_+)\rho(dx)},\tag{6.49}$$

where $\rho(dx)$ is the EMC $\{\xi_n; n \geq 0\}$ stationary distribution, $P(x,dy)$ are the EMC transition probabilities, $F_y(t)$ is DF of sojourn time in the state y.

For SMP with finite set of states this formula is given in [23].

To apply the formula (6.49) we need to know EMC $\{\xi_n; n \geq 0\}$ stationary distribution. The EMC transition probabilities are:

$$p_{11x}^{10y} = f_1(x-y),\ \ 0<y<x,\ \ p_{11x}^{20y} = f_1(x+y),\ \ y>0,\ \ p_{21x}^{20y} = f_2(x-y),$$
$$0<y<x\ \ p_{21x}^{10y} = f_2(x+y),\ \ y>0,\ \ P_{10x}^{11x}=1,\ \ P_{20x}^{21x}=1.$$

Let us introduce the notations $\rho_{10}(x)$, $\rho_{11}(x)$, $\rho_{20}(x)$, $\rho_{21}(x)$ of stationary distribution densities for the states $10x$, $11x$, $20x$, $21x$ correspondingly. The system of integral equations for the stationary distribution determination is as follows:

$$\begin{cases} \rho_{10}(x) = \int\limits_x^\infty f_1(y-x)\rho_{11}(y)\,dy + \int\limits_0^\infty f_2(x+y)\rho_{21}(y)\,dy, \\ \rho_{20}(x) = \int\limits_x^\infty f_2(y-x)\rho_{21}(y)\,dy + \int\limits_0^\infty f_1(x+y)\rho_{11}(y)\,dy, \\ \rho_{11}(x) = \rho_{10}(x),\ \ \rho_{21}(x) = \rho_{20}(x), \\ \int\limits_0^\infty \big(\rho_{11}(x)+\rho_{10}(x)+\rho_{21}(x)+\rho_{20}(x)\big)\,dx = 1. \end{cases}$$

The solution of this system is as follows:

$$\rho_{11}(x) = \rho_{10}(x) = \rho_0\bar{F}_2(x), \rho_{21}(x) = \rho_{20}(x) = \rho_0\bar{F}_1(x),$$

where the constant ρ_0 is obtained from the normalization requirement.

Let us find the expressions from the numerator and denominator of formula (6.49).

$$\int\limits_{E_-}\rho(dx)\int\limits_{E_+}\bar{F}_y(t)P(x,dy) = \rho_0\int\limits_0^\infty \bar{F}_2(x)\,\bar{\Phi}_1(x,t)\,dx =$$

$$\rho_0\Bigg(\int\limits_t^\infty \bar{F}_2(x)dx\bar{F}_1(t) + \int\limits_0^t \bar{F}_2(x)dx\bar{F}_1(t) +$$

$$+\int\limits_0^t \bar{F}_2(x)\,dx\int\limits_0^{t-x} f_1(x+y)\bar{G}_2(t-x-y)\,dy +$$

$$+\sum_{n=1}^\infty\int\limits_0^t \bar{F}_2(x)\,dx\int\limits_0^{t-x} f_1(x+y)F_2^{*(n)}(y)\big(G_2^{*(n)}(t-x-y)-G_2^{*(n+1)}(t-x-y)\big)dy\Bigg) =$$

$$= \rho_0\Bigg(E\alpha_2\bar{F}_1(t) + \int\limits_0^t \bar{F}_2(x)\,dx\int\limits_x^t f_1(s)\bar{G}_2(t-s)\,ds + \qquad (6.50)$$

$$+\sum_{n=1}^{\infty}\int_0^t \bar F_2(x)\,dx\int_x^t f_1(s)F_2^{*(n)}(s-x)\,ds\int_0^{t-s} g_0^{*(n)}(y)\bar G_2(t-s-y)\,dy\Bigg)=$$

$$=\rho_0\left(E\alpha_2\bar F_1(t)+\int_0^t \bar F_2(x)\,dx\int_x^t f_1(s)\bar G_2(t-s)\,ds+\right.$$

$$\left.+\sum_{n=1}^{\infty}\int_0^t \bar F_2(x)\,dx\int_x^t f_1(s)F_2^{*(n)}(s-x)\,dx\int_s^t g_2^{*(n)}(\sigma-s)\bar G_2(t-\sigma)\,d\sigma\right),$$

$$\int_{E_-} P(x,E_+)\rho(dx)=\rho_0 E\alpha_2. \tag{6.51}$$

By substituting (6.50) and (6.51) into (6.49), we get the expression for the function $\bar F_\theta(t)=1-F_\theta(t)$:

$$\bar F_\theta(t)=\frac{1}{E\alpha_2}\left(E\alpha_2\bar F_1(t)+\int_0^t \bar F_2(x)\,dx\int_x^t f_1(s)\bar G_2(t-s)\,ds+\right.$$
$$\left.+\sum_{n=1}^{\infty}\int_0^t \bar F_2(x)\,dx\int_x^t f_1(s)F_2^{*(n)}(s-x)\,ds\int_s^t g_2^{*(n)}(y-s)\bar G_2(t-y)\,dy\right). \tag{6.52}$$

Applying the formula (6.52), let us determine the mathematical expectation of RV θ.

$$E\theta=\int_0^{\infty}\bar F_\theta(t)\,dt=\frac{1}{E\alpha_2}\left(E\alpha_1 E\alpha_2+\int_0^{\infty}dt\int_0^t \bar F_2(x)\,dx\int_x^t f_1(s)\bar G_2(t-s)\,ds+\right.$$
$$\left.+\sum_{n=1}^{\infty}\int_0^{\infty}dt\int_0^t \bar F_2(x)\,dx\int_x^t f_1(s)F_2^{*(n)}(s-x)\,ds\int_s^t g_2^{*(n)}(y-s)\bar G_2(t-y)\,dy\right). \tag{6.53}$$

Let us transform the second and the third summand of the sum by changing the integration order subsequently.

$$\int_0^{\infty}dt\int_0^t \bar F_2(x)\,dx\int_x^t f_1(s)\bar G_2(t-s)\,ds=E\beta_2\int_0^{\infty}\bar F_2(x)\bar F_1(x)\,dx=E\beta_2\int_0^{\infty}f_1(s)\,ds\int_0^s \bar F_2(x)\,dx,$$
$$\sum_{n=1}^{\infty}\int_0^{\infty}dt\int_0^t \bar F_2(x)dx\int_x^t f_1(s)F_2^{*(n)}(s-x)\,ds\int_s^t g_2^{*(n)}(y-s)\bar G_2(t-y)\,dy= \tag{6.54}$$

$$=E\beta_2\int_0^{\infty}\bar F_2(x)\,dx\int_0^{\infty}f_1(x+y)\sum_{n=1}^{\infty}F_2^{*(n)}(y)dy=E\beta_2\int_0^{\infty}\bar F_2(x)\,dx\int_0^{\infty}f_1(x+y)H_2(y)\,dy=$$

$$= E\beta_2 \int_0^\infty f_1(s)\,ds \int_0^s H_2(s - x\overline{F}_2(x)\,dx,$$

where $H_2(x) = \sum_{n=1}^{\infty} F_2^{*(n)}(x)$ is the renewal function.

Taking into account (6.53) and (6.54) as well as the equality (1.13)

$$\int_0^s \overline{F}_2(x)\,dx + \int_0^s H_2(s-x)\overline{F}_2(x)\,dx = s, \tag{6.55}$$

we get

$$E\theta = \frac{1}{E\alpha_2}\left(E\alpha_1 E\alpha_2 + E\beta_2 \int_0^\infty f_1(s)\,ds \int_0^s \overline{F}_2(x)\,dx + \right.$$
$$\left. + E\beta_2 \int_0^\infty f_1(s)\,ds \int_0^s H_2(s-x)\overline{F}_2(x)\,dx \right) =$$
$$= \frac{E\alpha_1 \left(E\alpha_2 + E\beta_2\right)}{E\alpha_2}.$$

In such a way, in the case of nondepreciatory failures, TC productivity P_{nd} is determined in the following way:

$$P_{\text{nd}} = \frac{E\alpha_2}{E\alpha_1 \left(E\alpha_2 + E\beta_2\right)}. \tag{6.56}$$

Note, one can obtain the last formula by using strong law of large numbers [18], herewith, $P_{\text{nd}} = \lim_{t\to\infty} \frac{N(t)}{t}$, where $N(t)$ is the number of system transitions into the subset $E\_$ within time interval $(0,t]$.

Let us determine the variance of RV θ applying the formula

$$D\theta = 2\int_0^\infty t\overline{F}_\theta(t)\,dt - E^2\theta. \tag{6.57}$$

Let us consider the integral

$$\int_0^\infty t\overline{F}_\theta(t)\,dt = \int_0^\infty t\overline{F}_1(t)\,dt + \frac{1}{E\alpha_2}\left(\int_0^\infty t\,dt \int_0^t \overline{F}_2(x)\,dx \int_x^t f_1(s)\,\overline{G}_2(t-s)\,ds + \right.$$
$$\left. + \sum_{n=1}^{\infty} \int_0^\infty t\,dt \int_0^t \overline{F}_2(x)\,dx \int_x^t f_1(s)\overline{F}_2^{*(n)}(s-x)\,ds \int_s^t g_2^{*(n)}(y-s)\,\overline{G}_2(t-y)\,dy \right). \tag{6.58}$$

Let us transform the summands of this sum with regard to (6.55) and

$$\int_0^\infty t\, g^{*(n)}(t)\,dt = nE\beta_2 \qquad \int_0^\infty t\overline{F}_1(t)\,dt = \frac{1}{2}\left(D\alpha_1 + E^2\alpha_1\right),$$

$$\int_0^\infty t\,dt \int_0^t \overline{F}_2(x)\,dx \int_x^t f_1(s)\,\overline{G}_2(t-s)\,ds = \int_0^\infty \overline{F}_2(x)\,dx \int_x^\infty f_1(s)\,ds \int_0^\infty (s+y)\,\overline{G}_2(y)\,dy =$$

$$= E\beta_2 \int_0^\infty \overline{F}_2(x)\,dx \int_x^\infty s f_1(s)\,ds + \frac{1}{2}\left(D\beta_2 + E^2\beta_2\right) \int_0^\infty f_1(s)\,ds \int_0^s \overline{F}_2(x)\,dx =$$

$$= E\beta_2 \int_0^\infty s f_1(s)\,ds \int_0^s \overline{F}_2(x)\,dx + \frac{1}{2}\left(D\beta_2 + E^2\beta_2\right) \int_0^\infty f_1(s)\,ds \int_0^s \overline{F}_2(x)\,dx.$$

Next,

$$\sum_{n=1}^\infty \int_0^\infty t\,dt \int_0^t \overline{F}_2(x)\,dx \int_x^t f_1(s) F_2^{*(n)}(s-x)\,ds \int_s^t g_2^{*(n)}(y-s)\overline{G}_2(t-y)\,dy =$$

$$= \sum_{n=1}^\infty \int_0^\infty \overline{F}_2(x)\,dx \int_x^\infty f_1(s) F_2^{*(n)}(s-x)\,ds \int_s^\infty g_2^{*(n)}(y-s)\,dy \int_0^\infty (y+\sigma)\,\overline{G}_2(\sigma)\,d\sigma =$$

$$E\beta_2 \int_0^\infty \overline{F}_2(x)\,dx \int_x^\infty s f_1(s) H_2(s-x)\,ds +$$

$$+ E^2\beta_2 \sum_{n=1}^\infty n \int_0^\infty \overline{F}_1(s)\,ds \int_0^s f_2^{\bullet(n)}(s-x)\overline{F}_2(x)\,dx +$$

$$+ \frac{1}{2}\left(D\beta_2 + E^2\beta_2\right) \int_0^\infty f_1(s)\,ds \int_0^s H_2(s-x)\overline{F}_2(x)\,dx =$$

$$= E\beta_2 \int_0^\infty \overline{F}_2(x)\,dx \int_x^\infty s f_1(s) H_2(s-x)\,ds +$$

$$+ E^2\beta_2 \int_0^\infty \overline{F}_1(s)\,ds \sum_{n=1}^\infty n\left(F_2^{*(n)}(s) - F_2^{*(n+1)}(s)\right) +$$

$$+ \frac{1}{2}\left(D\beta_2 + E^2\beta_2\right) \int_0^\infty f_1(s)\,ds \int_0^s H_2(s-x)\overline{F}_2(x)\,dx =$$

$$= E\beta_2 \int_0^\infty \overline{F}_2(x)dx \int_x^\infty s f_1(s) H_2(s-x)ds + E^2\beta_2 \int_0^\infty \overline{F}_1(s)H_2(s)ds +$$

$$+\frac{1}{2}\left(D\beta_2 + E^2\beta_2\right) \int_0^\infty f_1(s)ds \int_0^s H_2(s-x)\overline{F}_2(x)dx.$$

The integral (6.58) takes the form:

$$\int_0^\infty t\overline{F}_\theta(t)dt = \frac{1}{2}\left(D\alpha_1 + E^2\alpha_1\right) + \frac{1}{E\alpha_2}\left(E\beta_2 \int_0^\infty s^2 f_1(s)ds + E^2\beta_2 \int_0^\infty \overline{F}_1(s)H_2(s)ds +\right.$$

$$\left. +\frac{1}{2}\left(D\beta_2 + E^2\beta_2 E\alpha_1\right)\right) = \frac{1}{2}\left(D\alpha_1 + E^2\alpha_1\right) + \frac{1}{E\alpha_2}\left(E\beta_2\left(D\alpha_1 + E^2\alpha_1\right) + \right. \tag{6.59}$$

$$\left. +\frac{1}{2}\left(D\beta_2 + E^2\beta_2 E\alpha_1\right) + E^2\beta_2 \int_0^\infty \overline{F}_1(s)H_2(s)ds\right).$$

Substituting (6.59) into the formula (6.57), we get

$$D\theta = \left(1 + \frac{2E\beta_2}{E\alpha_2}\right)D\alpha_1 + \frac{E\alpha_1}{E\alpha_2}\left(D\beta_2 + E^2\beta_2\left(1 - \frac{E\alpha_1}{E\alpha_2}\right)\right) + \tag{6.60}$$

$$+\frac{2E^2\beta_2}{E\alpha_1}\int_0^\infty \overline{F}_1(s)H_2(s)ds.$$

Let us write out formulas to define efficiency and variance of time of production service cycle for specific distributions of RVs α_1, α_2, β_2.

If RVs α_1, α_2, β_2 have exponential distributions with densities $f_1(t) = \lambda_1 e^{-\lambda_1 t}$, $f_2(t) = \lambda_2 e^{-\lambda_2 t}$, $g_2(t) = \mu e^{-\mu t}$ correspondingly, then from (6.56) and (6.60) we conclude:

$$P_{nd} = \frac{\lambda_1\mu}{\mu + \lambda_2}, D(\theta) = \frac{(\mu + \lambda_2)^2 + 2\lambda_1\lambda_2}{\lambda_1^2\lambda_2^2}. \tag{6.61}$$

Let us consider the case when RVs α_1, α_2, β_2 have Erlangian distributions with the densities

$$f_1(t) = \frac{\lambda_1^k t^{k-1} e^{-\lambda_1 t}}{(k-1)!}, f_2(t) = \frac{\lambda_2^m t^{m-1} e^{-\lambda_{21} t}}{(m-1)!}, g_2(t) = \frac{\mu^n t^{n-1} e^{-\mu t}}{(n-1)!}$$

correspondingly. Taking into account that

$$E\alpha_1 = \frac{k}{\lambda_1},\ D\alpha_1 = \frac{k}{\lambda_1^2}, E\alpha_2 = \frac{m}{\lambda_2},\ D\alpha_1 = \frac{m}{\lambda_2^2}, E\beta_2 = \frac{n}{\mu},\ D\beta = \frac{n}{\mu^2},$$

$$\overline{F_1}(x) = e^{-\lambda_1 t}\sum_{l=0}^{k-1}\frac{(\lambda_1 x)^l}{l!},\ H_2(x) = \frac{1}{m}\left(\lambda_2 x + \sum_{j=1}^{m-1}\frac{c^j}{1-c^j}\left(1-e^{-\lambda_2 x(1-c^j)}\right)\right),$$

$$c = e^{\frac{2\pi i}{m}}, i^2 = -1,$$

In accordance with formulas (6.56) and (6.60) we get

$$P_{nd} = \frac{m\lambda_1\mu}{k(\mu m + n\lambda_2)}, \tag{6.62}$$

$$D(\theta) = \frac{k\left((m\mu + n\lambda_2)^2 + \lambda_1\lambda_2 mn(n+1)\right)}{(m\lambda_1\mu)^2} +$$

$$+\frac{2n^2\lambda_2}{\mu^2 m^2\lambda_1}\sum_{l=0}^{k-1}\sum_{j=1}^{m-1}\frac{c^j}{1-c^j}\left(1-\left(\frac{\lambda_1}{\lambda_1+\lambda_2(1-c^j)}\right)^{l+1}\right).$$

We can prove TC productivity with depreciatory failures P_d looks like:

$$P_d = \frac{\int_0^\infty H_1(x) f_2(x)dx}{E\alpha_2 + E\beta_2}, \tag{6.63}$$

where $H_1(x) = \sum_{n=1}^{\infty} F_1^{*(n)}(x)$ is the renewal function.

In Table 6.3, the results of calculations by means of formulas (6.56), (6.63) are given. This is the case of RVs α_1, α_2, β_2 Erlangian distribution of the second order.

TABLE 6.3 TC Productivity in Cases of Nondepreciatory and Depreciatory Failures

$E\alpha_1$, h	$E\alpha_2$, h	$E\beta_2$, h	P_{nd}, unit/h (6.56)	P_d, unit/h (6.63)
0.05	2	0.2	18.18	18.00
		0.3	17.39	17.21
		0.4	16.66	16.50
	4	0.2	19.04	18.95
		0.3	18.60	18.51
		0.4	18.18	18.09
	6	0.2	19.35	19.29
		0.3	19.04	18.98
		0.4	18.74	18.68
0.1	2	0.2	9.09	8.90
		0.3	8.69	8.52
		0.4	8.33	8.16
	4	0.2	9.52	9.42
		0.3	9.30	9.21
		0.4	9.09	9.00
	6	0.2	9.68	9.61
		0.3	9.46	9.42
		0.4	9.37	9.31
0.2	2	0.2	4.55	4.36
		0.3	4.37	4.17
		0.4	4.16	4.00
	4	0.2	4.76	4.67
		0.3	4.65	4.56
		0.4	4.55	4.45
	6	0.2	4.83	4.77
		0.3	4.76	4.69
		0.4	4.69	4.62

Results of this section allow to carry out the analysis of influence of reliability of TC on its productivity.

Appendix A

The Solution of the System of Integral Equations (2.24)

Let us prove the system of equation (2.24) to have the solution determined by the formula (2.25).

Introduce the operators:

$$(A_r\varphi)(x)=\int_x^\infty r(y-x)\varphi(y)\,dy,\left(\bar{A}_r\varphi\right)(x)=\int_0^\infty r(x+y)\varphi(y)\,dy,$$

$$(A_v\varphi)(x)=\int_x^\infty v(y-x)\varphi(y)\,dy,\left(\bar{A}_v\varphi\right)(x)=\int_0^\infty v(x+y)\varphi(y)\,dy.$$

Then the system of equation (2.24) takes the following form:

$$\begin{cases}\rho_0=\rho(111)=\rho(222),\\ \rho(210x)=\rho_0A_rf+A_r\rho(211x),\\ \rho(211x)=A_v\rho(210x),\\ \rho(100x)=\bar{A}_v\rho(210x),\\ \rho(222)=\int_0^\infty\rho(100x)\,dx+\rho(200),\\ \rho(101x)=\rho_0\bar{A}_rf+\bar{A}_r\rho(211x),\\ \rho(200)=\int_0^\infty\rho(101x)\,dx,\\ 2\rho_0+\rho(200)+\int_0^\infty\left(\rho(210x)+\rho(211x)+\rho(100x)+\rho(101x)\right)dx=1.\end{cases}\tag{A.1}$$

The last equation of the system (A.1) – is a normalization requirement. By substituting the third equation of (A.1) in the second one, we get:

$$\rho(210x)=\rho_0A_rf+A_rA_v\rho(210x).\tag{A.2}$$

Let us denote

$$A_rA_v=A_\pi,\ \pi=r*v,\ \pi(x)=\int_0^x r(x-t)v(t)\,dt,\ \left(A_\pi\varphi\right)(x)=\int_x^\infty\pi(y-x)\varphi(y)\,dy\tag{A.3}$$

Semi-Markov Models. http://dx.doi.org/10.1016/B978-0-12-802212-2.00006-1
Copyright © 2015 Elsevier Inc. All rights reserved.

With regard to (A.3), the equation (A.2) takes the following form:

$$\rho(210x) = \rho_0 A_r f + A_\pi \rho(210x).$$

Then,

$$\rho(210x) = \rho_0 \left(I - A_\pi\right)^{-1} A_r f, \left(I - A_\pi\right)^{-1} = I + \sum_{n=1}^{\infty} A_\pi^n$$

$$\left(A_\pi^n \varphi\right)(x) = \int_x^\infty \pi^{*(n)}(y - x)\varphi(y)\,dy, \tag{A.4}$$

$$\rho(210x) = \rho_0 \int_x^\infty h^{(0)}(t - x) f(t)\,dt,$$

where $h^{(0)}(t) = \sum_{n=1}^{\infty} r * (r * v)^{*(n-1)}(t)$ is the density of the function of 0-renewation.

Substitute (A.4) in the third equation of the system (A.1):

$$\rho(211x) = \rho_0 A_v \left(I - A_\pi\right)^{-1} A_r f,$$

$$\rho(211x) = \rho_0 \int_x^\infty h^{(1)}(t - x) f(t)\,dt, \tag{A.5}$$

where $h^{(1)}(t) = \sum_{n=1}^{\infty} (r * v)^{*(n)}(t)$ is the density of the function of 1-renewation.

Substitute (A.4) in the fourth equation of (A.1):

$$\rho(100x) = \rho_0 \bar{A}_v \left(I - A_\pi\right)^{-1} A_r f = \rho_0 \int_0^\infty v(x + y)\,dy \int_y^\infty h^{(0)}(t - y) f(t)\,dt =$$

$$= \rho_0 \int_0^\infty f(t)\,dt \int_0^t v(x+y) h^{(0)}(t-y)\,dy = \rho_0 \int_0^\infty f(t)\,dt \int_0^t v(x+t-y) h^{(0)}(y)\,dy =$$

$$= \rho_0 \int_0^\infty f(t) v^{(0)}(t,x)\,dt,$$

where $v^{(0)}(t,x) = \int_0^t v(x + t - y) h^{(0)}(y)\,dy$ is the distribution density of the residual time of control.

In such a way,

$$\rho(100x) = \rho_0 \int_0^{\infty} f(t) v^{(0)}(t,x)\,dt. \tag{A.6}$$

Taking into account (A.5), with the help of the sixth equation of the system (A.1), we get:

$$\rho(101x) = \rho_0 \left(\bar{A}_r f + \bar{A}_r A_v \left(I - A_\pi \right)^{-1} A_r f \right) =$$

$$= \rho_0 \left(\int_0^{\infty} r(x+y) f(y)\,dy + \int_0^{\infty} r(x+y)\,dy \int_y^{\infty} h^{(1)}(t-y) f(t)\,dt \right) =$$

$$= \rho_0 \left(\int_0^{\infty} r(x+y) f(y)\,dy + \int_0^{\infty} f(t)\,dt \int_y^{\infty} r(x+t-y) h^{(1)}(y)\,dy \right) = \rho_0 \int_0^{\infty} f(t) v^{(1)}(t,x)\,dt,$$

$$\text{where } v^{(1)}(t,x) = r(t+x) + \int_0^{t} r(x+t-y) h_1(y)\,dy \tag{A.7}$$

is the density of the direct residual time till the control beginning.

$$\text{Concequently, } \rho(101x) = \rho_0 \int_0^{\infty} f(t) v^{(1)}(t,x)\,dt. \tag{A.8}$$

Let us integrate the expression (A.7):

$$\begin{aligned} \int_0^{\infty} v^{(1)}(t,x)\,dx &= \int_0^{\infty} r(t+x)\,dx + \int_0^{\infty} dx \int_0^{t} r(x+t-y) h^{(1)}(y)\,dy = \\ &= \bar{R}(t) + \int_0^{t} \bar{R}(t-y) h^{(1)}(y)\,dy = \bar{V}^{(1)}(t,0) = \\ &= 1 - \bar{V}^{(0)}(t,0) = 1 - H^{(0)}(t) + H^{(1)}(t) = \tilde{H}^{(1)}(t) - H^{(0)}(t). \end{aligned} \tag{A.9}$$

Applying the seventh equation of the system (A.1), with regard to the equality (A.9), we have:

$$\rho(200) = \rho_0 \int_0^{\infty} \left(\tilde{H}^{(1)}(t) - H^{(0)}(t) \right) f(t)\,dt. \tag{A.10}$$

Then, due to (A.4), (A.5), (A.6), (A.8), and (A.10), the solution of the system (2.24) is defined by the formula (2.25).

The value of ρ_0 can be found by means of normalization requirement included in (2.25).

$$3\rho_0 + \rho_0 \int_0^{\infty} f(t)\,dt \int_0^{\infty} v^{(1)}(t,x)\,dx + \rho_0 \int_0^{\infty} \bar{F}(t) h^{(0)}(t)\,dt + \rho_0 \int_0^{\infty} \bar{F}(t) h^{(1)}(t)\,dt = 1. \tag{A.11}$$

Due to the equality (A.9), the expression (A.11) takes the following form:

$$3\rho_0+\rho_0\int_0^{\infty}\left(1-H^{(0)}(t)+H^{(1)}(t)\right)f(t)dt+\rho_0\int_0^{\infty}\bar{F}(t)h^{(0)}(t)dt+\rho_0\int_0^{\infty}\bar{F}(t)h^{(1)}(t)dt=1.$$

Next, $3\rho_0+\rho_0-\rho_0\int_0^{\infty}\bar{F}(t)h^{(0)}(t)dt+\rho_0\int_0^{\infty}\bar{F}(t)h^{(1)}(t)dt+$

$$+\rho_0\int_0^{\infty}\bar{F}(t)h^{(0)}(t)dt+\rho_0\int_0^{\infty}\bar{F}(t)h^{(1)}(t)dt=1{,}4\rho_0+2\rho_0\int_0^{\infty}\bar{F}(t)h_1(t)dt=1.$$

Consequently,

$$\rho_0=\left(4+2\int_0^{\infty}H^{(1)}(t)f(t)dt\right)^{-1}=\frac{1}{2}\left(1+\int_0^{\infty}\tilde{H}^{(1)}(t)f(t)dt\right)^{-1}.$$

Identity (2.42) proof

Let us prove the identity (2.42):

$$\int_0^{\infty}f(t)dt\int_0^{t}h^{(0)}(t-x)dx\int_0^{x}\bar{V}(y)dy+\int_0^{\infty}f(t)dt\int_0^{\infty}\bar{V}^{(0)}(t,x)dx=E\gamma\int_0^{\infty}H^{(0)}(t)f(t)dt. \quad (A.12)$$

Consider the integral $\int_0^{\infty}\bar{V}^{(0)}(t,x)dx$.

$$\int_0^{\infty}\bar{V}^{(0)}(t,x)dx=\int_0^{\infty}dx\int_0^{t}\bar{V}(t+x-s)h^{(0)}(s)ds=\int_0^{t}h^{(0)}(s)ds\int_{t-s}^{\infty}\bar{V}(y)dy=$$
$$=\int_0^{t}h^{(0)}(t-x)dx\int_x^{\infty}\bar{V}(y)dy. \quad (A.13)$$

With regard to (A.13), the expression (A.12) is as follows:

$$\int_0^{\infty}f(t)dt\int_0^{t}h^{(0)}(t-x)dx\int_0^{x}\bar{V}(y)dy+\int_0^{\infty}f(t)dt\int_0^{\infty}\bar{V}^{(0)}(t,x)dx=$$

$$=\int_0^{\infty}f(t)dt\int_0^{t}h^{(0)}(t-x)dx\int_0^{x}\bar{V}(y)dy+\int_0^{\infty}f(t)dt\int_0^{\infty}h^{(0)}(t-x)dx\int_x^{\infty}\bar{V}(y)dy=$$

$$=\int_0^{\infty}f(t)dt\int_0^{t}h^{(0)}(t-x)dx\left(\int_0^{x}\bar{V}(y)dy+\int_x^{\infty}\bar{V}(y)dy\right)=E\gamma\int_0^{\infty}H^{(0)}(t)f(t)dt.$$

Appendix B

The Solution of the System of Integral Equations (2.74)

To solve the system (2.74), let us substitute its third equation into the second one:

$$\rho(2\hat{1}0x) = \rho_0 \int_0^{\infty} f(x+t)r(t)\,dt + \overline{p}_1 \int_x^{\infty} \rho(2\hat{1}0y)r(y-x)\,dy. \tag{B.1}$$

Denoting $\rho(2\hat{1}0x) = \varphi_1(x)$, we get the equation (B.1) in the following form:

$$\varphi_1(x) = \rho_0 \int_x^{\infty} f(t)r(t-x)\,dt + \overline{p}_1 \int_x^{\infty} \varphi_1(y)r(y-x)\,dy. \tag{B.2}$$

Introduce the function $\tilde{r}(x) = \overline{p}_1 r(x)$ and the operator

$$\left(A_{\tilde{r}}\varphi\right)(x) = \int_x^{\infty} \tilde{r}(y-x)\varphi(y)\,dy,$$

then the equation (B.2) is as follows:

$$\varphi_1 = A_{\tilde{r}}\varphi_1 + \frac{\rho_0}{\overline{p}_1} A_{\tilde{r}} f, (I - A_{\tilde{r}})\varphi_1 = \frac{\rho_0}{\overline{p}_1} A_{\tilde{r}} f,$$

hence,

$$\varphi_1 = (I - A_{\tilde{r}})^{-1}\left[\frac{\rho_0}{\overline{p}_1} A_{\tilde{r}} f\right] = \sum_{n=0}^{\infty} A_{\tilde{r}}^n \left[\frac{\rho_0}{\overline{p}_1} A_{\tilde{r}} f\right] = \frac{\rho_0}{\overline{p}_1} \sum_{n=1}^{\infty} A_{\tilde{r}}^n f.$$

Let us write out the n-th power of operator $A_{\tilde{r}}$.

$$\left(A_{\tilde{r}}^2\varphi\right)(x) = \int_x^{\infty} \tilde{r}(y-x)\,dy \int_y^{\infty} \tilde{r}(z-y)\varphi(z)\,dz = \int_x^{\infty} \varphi(z)\,dz \int_x^{z} \tilde{r}(y-x)\tilde{r}(z-y)\,dy =$$

$$= \int_x^{\infty} \varphi(z)\,dz \int_x^{z-z} \tilde{r}(y)\tilde{r}(z-x-y)\,dy = \int_x^{\infty} \tilde{r}^{*(2)}(z-x)\varphi(z)\,dz,$$

Semi-Markov Models. http://dx.doi.org/10.1016/B978-0-12-802212-2.00007-3
Copyright © 2015 Elsevier Inc. All rights reserved.

$$\left(A_{\tilde{r}}^n\varphi\right)(x)=\int_x^{\infty}\tilde{r}^{*(n)}(y-x)\varphi(y)\,dy,\ \left(\sum_{n=1}^{\infty}A_{\tilde{r}}^n\varphi\right)(x)=\int_x^{\infty}h_{\tilde{r}}(t-x)\varphi(t)\,dt,$$

where $h_{\tilde{r}}(t)=\sum_{n=1}^{\infty}\tilde{r}^{*(n)}(t)$, $\tilde{r}^{*(n)}(t)$ is n-th fold convolution of the function $\tilde{r}(t)$.
Then,

$$\rho(2\hat{1}0x)=\varphi_1(x)=\rho_0\frac{1}{\overline{p}_1}\int_x^{\infty}h_{\tilde{r}}(y-x)f(y)dy=\frac{\rho_0}{\overline{p}_1}\int_0^{\infty}h_{\tilde{r}}(y)f(x+y)dy. \quad \text{(B.3)}$$

With regard to (B.3), we get the rest of sought functions.

$$\rho(211x)=\rho_0\int_0^{\infty}h_{\tilde{r}}(y)f(x+y)dy,$$

$$\rho(101x)=\rho_0\left(\int_0^{\infty}r(x+t)f(t)dt+\int_0^{\infty}r(x+y)dy\int_0^{\infty}h_{\tilde{r}}(z)f(y+z)dz\right)=$$

$$=\rho_0\left(\int_0^{\infty}r(x+z)f(z)dz+\int_0^{\infty}r(x+y)dy\int_y^{\infty}h_{\tilde{r}}(z-y)f(z)dz\right)= \quad \text{(B.4)}$$

$$=\rho_0\left(\int_0^{\infty}r(x+z)f(z)dz+\int_0^{\infty}f(z)dz\int_0^{z}r(x+y)h_{\tilde{r}}(z-y)dy\right)=$$

$$=\frac{\rho_0}{\overline{p}_1}\int_0^{\infty}f(z)dz\left(\tilde{r}(x+z)+\int_0^{z}\tilde{r}(x+y)h_{\tilde{r}}(z-y)dy\right)=\frac{\rho_0}{\overline{p}_1}\int_0^{\infty}f(z)\tilde{v}_r(z,x)dz,$$

where $\tilde{v}_r(z,x)=\tilde{r}(x+z)+\int_0^{z}\tilde{r}(x+y)h_{\tilde{r}}(z-y)dy$,

$$\rho(2\bar{1}2)=p_1\int_0^{\infty}\rho(2\hat{1}0x)dx=\rho_0\frac{p_1}{\overline{p}_1}\int_0^{\infty}dx\int_0^{\infty}h_{\tilde{r}}(y)f(x+y)dy=\rho_0\frac{p_1}{\overline{p}_1}\int_0^{\infty}\overline{F}(y)h_{\tilde{r}}(y)dy.$$

Substitute (B.4) in the fifth equation of the system (2.74):

$$\rho(2\hat{0}0)=\int_0^{\infty}\rho(101x)\,dx+p_0\rho(2\hat{0}0),\ \overline{p}_0\rho(2\hat{0}0)=\int_0^{\infty}\rho(101x)\,dx,$$

$$\rho(2\hat{0}0)=\frac{1}{\overline{p}_0}\int_0^{\infty}\rho(101x)dx=\rho_0\frac{1}{\overline{p}_0\overline{p}_1}\int_0^{\infty}dx\int_0^{\infty}\tilde{v}(z,x)f(z)dz= \quad \text{(B.5)}$$

$$=\rho_0\frac{1}{\overline{p}_0\overline{p}_1}\int_0^{\infty}f(z)dz\int_0^{\infty}\tilde{v}(z,x)dx=\rho_0\frac{1}{\overline{p}_0}\left(1-\frac{p_1}{\overline{p}_1}\int_0^{\infty}H_{\tilde{r}}(z)f(z)dz\right).$$

Here, $H_{\bar{r}}(t)=\sum_{n=0}^{\infty}\tilde{R}^{*(n)}(t)$, where $\tilde{R}^{*(n)}(t)$ *is n-th* fold convolution of the function $\tilde{R}(t)$, $\tilde{R}(t)=\bar{p}_1 R(t)$.

To make the above transformation the following equality was applied:

$$\int_0^{\infty}\tilde{v}_r(z,x)dx=\bar{p}_1-p_1 H_{\bar{r}}(z). \tag{B.6}$$

Its proof is given hereafter.

Applying the formula (B.5), we obtain $\rho(2\bar{0}1)$ and $\rho(222)$:

$$\rho(2\bar{0}1)=p_0\rho(2\hat{0}0)=\rho_0\frac{p_0}{\bar{p}_0}\left(1-\frac{p_1}{\bar{p}_1}\int_0^{\infty}H_{\bar{r}}(z)f(z)dz\right),$$

$$\rho(222)=\bar{p}_0\rho(2\hat{0}0)=\rho_0\left(1-\frac{p_1}{\bar{p}_1}\int_0^{\infty}H_{\bar{r}}(z)f(z)dz\right).$$

In such a way, the solution of the system (2.74) is given by the formula (2.75).

The value of constant ρ_0 can be found by means of the normalization requirement.

Let us prove the equality (B.6):

$$\int_0^{\infty}\tilde{v}_r(z,x)dx=\int_0^{\infty}\bar{r}(x+z)dx+\int_0^{\infty}dx\int_0^{z}\bar{r}(x+y)h_{\bar{r}}(z-y)dy=$$

$$=\bar{p}_1\bar{R}(z)+\int_0^{z}h_{\bar{r}}(z-y)dy\int_0^{\infty}\bar{r}(x+y)dx=\bar{p}_1\bar{R}(z)+\int_0^{z}h_{\bar{r}}(z-y)\bar{p}_1\bar{R}(y)dy= \tag{B.7}$$

$$=\bar{p}_1\bar{R}(z)+\bar{p}_1\int_0^{z}\bar{R}(z-y)h_{\bar{r}}(y)dy=\bar{p}_1\bar{R}(z)+\bar{p}_1H_{\bar{r}}(z)-\int_0^{z}H_{\bar{r}}(z-y)\bar{p}_1r(y)dy.$$

To convert the last summand of (B.7), let us consider the function $h_{\bar{r}}(t)$. The following equality is true:

$$h_{\bar{r}}(t)=\bar{r}(t)+\int_0^{t}h_{\bar{r}}(t-x)\bar{r}(x)dx. \tag{B.8}$$

By integrating the expression (B.8), we get:

$$\int_0^{z}h_{\bar{r}}(t)dt=\int_0^{z}\bar{r}(t)dt+\int_0^{z}dt\int_0^{t}h_{\bar{r}}(t-x)\bar{r}(x)dx;$$

$$H_{\bar{r}}(z) = \bar{p}_1 R(z) + \int_0^z \tilde{r}(x)\,dx \int_x^z h_{\bar{r}}(t-x)\,dt = \bar{p}_1 R(z) + \int_0^z \tilde{r}(x) H_{\bar{r}}(z-x)\,dx =$$

$$= \bar{p}_1 R(z) + \int_0^z \bar{p}_1 r(x) H_{\bar{r}}(z-x)\,dx.$$

Then,

$$\int_0^z H_{\bar{r}}(z-x)\bar{p}_1 r(x)\,dx = H_{\bar{r}}(z) - \bar{p}_1 R(z). \tag{B.9}$$

With regard to (B.9), the formula (B.7) will take the following form:

$$\int_0^\infty \tilde{v}_r(z,x)\,dx = \bar{p}_1 \bar{R}(z) + \bar{p}_1 H_{\bar{r}}(z) - \left(H_{\bar{r}}(z) - \bar{p}_1 R(z)\right) =$$

$$= \bar{p}_1 \bar{R}(z) + \bar{p}_1 R(z) + \bar{p}_1 H_{\bar{r}}(z) - H_{\bar{r}}(z) = \bar{p}_1 - p_1 H_{\bar{r}}(z). \tag{B.10}$$

Let us prove the equality:

$$\int_0^z \bar{R}(z-t) h_{\bar{r}}(t)\,dt = R(z) - \frac{p_1}{\bar{p}_1} H_{\bar{r}}(z),$$

which was applied in the formula (2.77).

$$\int_0^z \bar{R}(z-t) h_{\bar{r}}(t)\,dt = \int_0^z \bar{R}(t) h_{\bar{r}}(z-t)\,dt = H_{\bar{r}}(z) - \frac{1}{\bar{p}_1}\int_0^z H_{\bar{r}}(z-t)\tilde{r}(t)\,dt. \tag{B.11}$$

Taking into account (B.9), we can write (B.11) in the following form:

$$\int_0^z \bar{R}(z-t) h_{\bar{r}}(t)\,dt = H_{\bar{r}}(z) - \frac{1}{\bar{p}_1}\left[H_{\bar{r}}(z) - \bar{p}_1 R(z)\right] = H_{\bar{r}}(z) - \frac{1}{\bar{p}_1} H_{\bar{r}}(z) + R(z) =$$

$$= H_{\bar{r}}(z)\left(1 - \frac{1}{\bar{p}_1}\right) + R(z) = R(z) - \frac{p_1}{\bar{p}_1} H_{\bar{r}}(z). \tag{B.12}$$

Let us obtain the value of the integral $\int_0^\infty x\tilde{v}(z,x)\,dx$ from the formula (2.78).

As $\tilde{v}(z,x) = \tilde{r}(z+x) + \int_0^z \tilde{r}(x+z-y) h_{\bar{r}}(y)\,dy$,

$$\int_0^\infty x\tilde{v}(z,x)\,dx = \int_0^\infty x\tilde{r}(z+x)\,dx + \int_0^\infty x\,dx \int_0^z \tilde{r}(x+z-y) h_{\bar{r}}(y)\,dy. \tag{B.13}$$

Integrating the first summand of (B.13) by parts, we have:

$$\int_0^\infty x\tilde{r}(z+x)\,dx = \bar{p}_1\int_z^\infty \bar{R}(x)\,dx. \tag{B.14}$$

Consider the second summand of the expression (B.13):

$$\begin{aligned}
&\int_0^\infty x\,dx\int_0^z \tilde{r}(x+z-y)h_{\tilde{r}}(y)\,dy = \int_0^z h_{\tilde{r}}(y)\,dy\int_0^\infty x\tilde{r}(x+z-y)\,dx = \\
&= \bar{p}_1\int_0^z h_{\tilde{r}}(y)\,dy\int_0^\infty \bar{R}(z+x-y)\,dx = \bar{p}_1\int_0^z h_{\tilde{r}}(y)\,dy\int_{z-y}^\infty \bar{R}(x)\,dx = \\
&= \bar{p}_1\int_0^z h_{\tilde{r}}(y)\,dy\int_0^\infty \bar{R}(x)\,dx - \bar{p}_1\int_0^z h_{\tilde{r}}(y)\,dy\int_0^{z-y} \bar{R}(x)\,dx = \\
&= \bar{p}_1 E\gamma\, H_{\tilde{r}}(z) - \bar{p}_1\int_0^z h_{\tilde{r}}(y)\,dy\int_0^{z-y} \bar{R}(x)\,dx.
\end{aligned} \tag{B.15}$$

Consider the second summand of the expression (B.15).

$$\begin{aligned}
&\bar{p}_1\int_0^z h_{\tilde{r}}(y)\,dy\int_0^{z-y} \bar{R}(x)\,dx = \bar{p}_1\int_0^z h_{\tilde{r}}(y)\,dy\int_y^z \bar{R}(x-y)\,dx = \\
&= \bar{p}_1\int_0^z dx\int_0^x \bar{R}(x-y)h_{\tilde{r}}(y)\,dy = \bar{p}_1\int_0^z dx\int_0^x \bar{R}(y)h_{\tilde{r}}(x-y)\,dy = \\
&= \int_0^z dx\left(\bar{p}_1 H_{\tilde{r}}(x) - \int_0^x H_{\tilde{r}}(x-y)\tilde{r}(y)\,dy\right) = \int_0^z dx\left(\bar{p}_1 H_{\tilde{r}}(x) - \left(H_{\tilde{r}}(x) - \bar{p}_1 R(x)\right)\right) = \\
&= \int_0^z\left(-p_1 H_{\tilde{r}}(x) + \bar{p}_1 R(x)\right)dx = -\,p_1\int_0^z H_{\tilde{r}}(x)\,dx + \bar{p}_1 z - \bar{p}_1\int_0^z \bar{R}(x)\,dx.
\end{aligned} \tag{B.16}$$

With regard to (B.14), (B.15), (B.16), the expression (B.13) takes the form:

$$\begin{aligned}
&\int_0^\infty x\tilde{v}(z,x)\,dx = \bar{p}_1\int_z^\infty \bar{R}(x)\,dx + \bar{p}_1 E\gamma\, H_{\tilde{r}}(z) + p_1\int_0^z H_{\tilde{r}}(x)\,dx - \bar{p}_1 z + \bar{p}_1\int_0^z \bar{R}(x)\,dx = \\
&\qquad = \bar{p}_1 E\gamma + \bar{p}_1 E\gamma H_{\tilde{r}}(z) + p_1\int_0^z H_{\tilde{r}}(x)\,dx - \bar{p}_1 z.
\end{aligned}$$

Appendix C

The Solution of the System of Integral Equation (3.6)

Let us prove the formula (3.9) to determine the solution of (3.6).

Introduce the operator: $(A_r\varphi)(x_1,x_2)=\int\limits_0^\infty \varphi(x_1+t,x_2+t)r(t)\,dt,$

then the second equation of the system (3.6) can be rewritten as follows:

$$\varphi_1=\rho_0 A_r[f_1(x_1)f_2(x_2)]+A_r\varphi_1+A_r[f_1(x_1)\varphi_4(x_2)]+A_r[\varphi_5(x_1)f_2(x_2)],$$

then,

$$(I-A_r)\varphi_1=\rho_0 A_r[f_1(x_1)f_2(x_2)]+A_r[f_1(x_1)\varphi_4(x_2)]+A_r[\varphi_5(x_1)f_2(x_2)],$$

$$\varphi_1=\rho_0(I-A_r)^{-1}A_r[f_1(x_1)f_2(x_2)]+(I-A_r)^{-1}A_r[f_1(x_1)\varphi_4(x_2)]+$$

$$+(I-A_r)^{-1}A_r[\varphi_5(x_1)f_2(x_2)],$$

$$(I-A_r)^{-1}=I+\sum_{n=1}^{\infty}A_r^n,\left(A_r^n\varphi\right)(x_1,x_2)=\int\limits_0^\infty \varphi(x_1+t,x_2+t)r^{*(n)}(t)\,dt,$$

$$\left(\sum_{n=1}^{\infty}A_r^n\right)\varphi=\int\limits_0^\infty \varphi(x_1+t,x_2+t)h_r(t)\,dt$$

where $h_r(t)=\sum_{n-1}^{\infty}r^{*(n)}(t)$ is the density of the renewal function $H(t)$ of the renewal process generated by RV δ.

Consequently,

$$\varphi_1(x_1,x_2)=\rho_0\int\limits_0^\infty f_1(x_1+y)f_2(x_2+y)h_r(y)dy+$$

$$+\int\limits_0^\infty \varphi_4(x_2+t)f_1(x_1+t)h_r(t)\,dt+\int\limits_0^\infty \varphi_5(x_1+t)f_2(x_2+t)h_r(t)\,dt. \qquad \text{(C.1)}$$

Semi-Markov Models. http://dx.doi.org/10.1016/B978-0-12-802212-2.00008-5
Copyright © 2015 Elsevier Inc. All rights reserved.

By substituting $\varphi_1(x_1, x_2)$ into the third equation of (3.6), we get:

$$\varphi_2(x,z) = \rho_0 \int_0^\infty f_1(t) f_2(x+t) r(z+t)\,dt + \int_0^\infty \varphi_4(x+t) f_1(t) r(z+t)\,dt$$

$$+\rho_0 \int_0^\infty r(z+t)\,dt \int_0^\infty f_1(t+y) f_2(x+t+y) h_r(y)\,dy +$$

$$+\int_0^\infty r(z+t)\,dt \int_0^\infty \varphi_4(x+t+y) f_1(t+y) h_r(y)\,dy -$$

$$+\int_0^\infty r(z+t)\,dt \int_0^\infty \varphi_5(t+y) f_2(x+t+y) h_r(y)\,dy. \tag{C.2}$$

Transform the equation (C.2).

$$\int_0^\infty f_1(t) f_2(x+t) r(z+t)\,dt + \int_0^\infty r(z+t)\,dt \int_0^\infty f_1(t+y) f_2(x+t+y) h_r(y)\,dy =$$

$$= \int_0^\infty f_1(t) f_2(x+t) r(z+t)\,dt + \int_0^\infty f_1(y) f_2(x+y)\,dy \int_0^y r(z+t) h_r(y-t)\,dt =$$

$$= \int_0^\infty f_1(y) f_2(x+y) v_r(y,z)\,dy,$$

where $v_r(y,z) = r(y+x) + \int_0^y r(y+z-s) h_r(s)\,ds$ is the distribution density of the direct residual time for the renewal process generated by RV δ.

Similarly,

$$\int_0^\infty \varphi_4(x+t) f_1(t) r(z+t)\,dt + \int_0^\infty r(z+t)\,dt \int_0^\infty \varphi_4(x+t+y) f_1(t+y) h_r(y)\,dy =$$

$$= \int_0^\infty f_1(y) \varphi_4(x+y) v_r(y,z)\,dy,$$

$$\int_0^\infty \varphi_5(t) f_2(x+t) r(z+t)\,dt + \int_0^\infty r(z+t)\,dt \int_0^\infty \varphi_5(t+y) f_2(x+t+y) h_r(y)\,dy =$$

$$= \int_0^\infty \varphi_5(y) f_2(x+y) v_r(y,z)\,dy.$$

The equation (C.2) takes the form:

$$\varphi_2(x,z) = \rho_0 \int_0^\infty f_1(y) f_2(x+y) v_r(y,z) dy +$$

$$+\int_0^\infty \varphi_4(x+y) f_1(y) v_r(y,z)\, dy + \int_0^\infty \varphi_5(y) f_2(x+y) v_r(y,z)\, dy. \tag{C.3}$$

In the same way, substituting $\varphi_1(x_1, x_2)$ into the fourth equation of the system (3.6) and making necessary transformations, we have:

$$\varphi_3(x,z) = \rho_0 \int_0^\infty f_2(y) f_1(x+y) v_r(y,z)\, dy +$$

$$+\int_0^\infty \varphi_5(x+y) f_2(y) v_r(y,z)\, dy + \int_0^\infty \varphi_4(y) f_1(x+y) v_r(y,z)\, dy. \tag{C.4}$$

Substitution of $\varphi_2(x,z)$ into the fifth equation of the system (3.6) results in:

$$\varphi_4(x) = \int_0^\infty \varphi_2(x+t,t)\, dt = \rho_0 \int_0^\infty dt \int_0^\infty f_1(y) f_2(x+t+y) v_r(y,t)\, dy +$$

$$+\int_0^\infty dt \int_0^\infty \varphi_5(y) f_2(x+t+y) v_r(y,t)\, dy + \int_0^\infty dt \int_0^\infty \varphi_4(x+t+y) f_1(y) v_r(y,t)\, dy =$$

$$= \rho_0 \int_0^\infty f_2(x+y)\, dy \int_0^y f_1(t) v_r(t, y-t) dt +$$

$$+\int_0^\infty f_2(x+y)\, dy \int_0^y \varphi_5(t) v_r(t, y-t)\, dt + \int_0^\infty \varphi_4(x+y)\, dy \int_0^y f_1(t) v_r(t, y-t)\, dt.$$

And we get analogous equation by substituting $\varphi_3(x,z)$ in the sixth equation of the system (3.6).

In such a way, we get the following system of equations:

$$\begin{cases} \varphi_4(x) = \rho_0 \int\limits_0^\infty f_2(x+y)\,dy \int\limits_0^y f_1(t)v_r(t,y-t)\,dt + \\ + \int\limits_0^\infty f_2(x+y)\,dy \int\limits_0^y \varphi_5(t)v_r(t,y-t)\,dt + \int\limits_0^\infty \varphi_4(x+y)\,dy \int\limits_0^y f_1(t)v_r(t,y-t)\,dt, \\ \varphi_5(x) = \rho_0 \int\limits_0^\infty f_1(x+y)\,dy \int\limits_0^y f_2(t)v_r(t,y-t)\,dt + \\ + \int\limits_0^\infty \varphi_5(x+y)\,dy \int\limits_0^y f_2(t)v_r(t,y-t)\,dt + \int\limits_0^\infty f_1(x+y)\,dy \int\limits_0^y \varphi_4(t)v_r(t,y-t)\,dt. \end{cases} \tag{C.5}$$

Introduce the functions: $\tilde{\gamma}_i(y) = \int\limits_0^y f_i(t)v_r(t,y-t)dt,\ i=1,\ 2$ and operators

$$(\Gamma_i\varphi)(x) = \int\limits_0^\infty \varphi(x+y)\tilde{\gamma}_i(y)\,dy,\ i=1,2.$$

One should note that the functions $\tilde{\gamma}_i(y)$ should be distribution densities and which are given as:

$$\int\limits_0^\infty \tilde{\gamma}_i(y) = \int\limits_0^\infty dy \int\limits_0^y f_i(t)v_r(t,y-t)\,dt = \int\limits_0^\infty f_i(t)\,dt \int\limits_t^\infty v_r(t,y-t)\,dy$$
$$= \int\limits_0^\infty f_i(t)\,dt \int\limits_0^\infty v_r(t,y)\,dy = 1.$$

Then the system (C.5) can be rewritten as follows:

$$\begin{cases} \varphi_4(x) = \rho_0\,(\Gamma_1 f_2)(x) + (\Gamma_1\varphi_4)(x) + \int\limits_0^\infty f_2(x+y)\,dy \int\limits_0^y \varphi_5(t)v_r(t,y-t)\,dt, \\ \varphi_5(x) = \rho_0\,(\Gamma_2 f_1)(x) + (\Gamma_2\varphi_5)(x) + \int\limits_0^\infty f_1(x+y)\,dy \int\limits_0^y \varphi_4(t)v_r(t,y-t)\,dt. \end{cases} \tag{C.6}$$

Let us consider the first equation of the system (C.6):

$$(I-\Gamma_1)\varphi_4(x) = \rho_0\,(\Gamma_1 f_2)(x) + \int\limits_0^\infty f_2(x+y)\,dy \int\limits_0^y \varphi_5(t)v_r(t,y-t)\,dt,$$

$$\varphi_4(x) = (I-\Gamma_1)^{-1}\left[\rho_0\,(\Gamma_1 f_2)(x) + \int\limits_0^\infty f_2(x+y)\,dy \int\limits_0^y \varphi_5(t)v_\kappa(t,y-t)\,dt\right],$$

$$(I-\Gamma_1)^{-1} = I + \sum_{n=1}^{\infty} \Gamma_1^n, \left(\Gamma_1^n\varphi\right)(x) = \int\limits_0^\infty \varphi(x+t)\tilde{\gamma}_1^{*(n)}(t)\,dt,$$

$$\left(\sum_{n=1}^{\infty}\Gamma_1^n\right)\varphi(x)=\int_0^{\infty}\varphi(x+t)h_1(t)\,dt,$$

where $h_1(t)=\sum_{n=1}^{\infty}\tilde{\gamma}_1^{*(n)}(t)$ is the density of the renewal function of the renewal process generated by RV with density $\tilde{\gamma}_1(t)$.

Consequently,

$$\varphi_4(x)=\rho_0\int_0^{\infty}f_2(x+y)h_1(y)\,dy+\int_0^{\infty}f_2(x+y)\,dy\int_0^{y}\varphi_5(t)v_r(t,y-t)\,dt+$$

$$+\int_0^{\infty}h_1(y)\,dy\int_0^{\infty}f_2(x+y+z)\,dz\int_0^{z}\varphi_5(t)v_r(t,z-t)\,dt. \qquad \text{(C.7)}$$

Let us transform the second summand of the right-hand side of (C.7):

$$\int_0^{\infty}f_2(x+y)\,dy\int_0^{y}\varphi_5(t)v_r(t,y-t)\,dt=\int_0^{\infty}\varphi_5(t)\,dt\int_0^{\infty}f_2(x+t+y)v_r(t,y)\,dy.$$

Introduce the function:

$$\beta_2(x,t)=\int_0^{\infty}f_2(x+y+t)v_r(t,y)\,dy. \qquad \text{(C.8)}$$

The third summand of the right-hand side of the equation (C.7) can be rewritten as follows:

$$\int_0^{\infty}h_1(y)\,dy\int_0^{\infty}f_2(x+y+z)\,dz\int_0^{z}\varphi_5(t)v_\kappa(t,z-t)\,dt=\int_0^{\infty}\varphi_5(t)\,dt\int_0^{\infty}\beta_2(x+y,t)h_1(y)\,dy.$$

Consequently, the first equation of the system (C.6) takes the form:

$$\varphi_4(x)=\rho_0\int_0^{\infty}f_2(x+y)h_1(y)\,dy+\int_0^{\infty}\varphi_5(t)\beta_2(x,t)\,dt+\int_0^{\infty}\varphi_5(t)\,dt\int_0^{\infty}\beta_2(x+y,t)h_1(y)\,dy.$$

In the same manner, the second equation of the system (C.6) can be simplified:

$$\varphi_5(x)=\rho_0\int_0^{\infty}f_1(x+y)h_2(y)\,dy+\int_0^{\infty}\varphi_4(t)\beta_1(x,t)\,dt+\int_0^{\infty}\varphi_4(t)\,dt\int_0^{\infty}\beta_1(x+y,t)h_2(y)\,dy,$$

where

$$\beta_1(x,t)=\int_0^{\infty}f_1(x+t+y)v_r(t,y)\,dy. \qquad \text{(C.9)}$$

Introduce the functions:

$$\gamma_1(x,t)=\beta_1(x,t)+\int_0^\infty \beta_1(x+y,t)h_2(y)\,dy,$$

$$\gamma_2(x,t)=\beta_2(x,t)+\int_0^\infty \beta_2(x+y,t)h_1(y)\,dy. \tag{C.10}$$

We get the following system of equations:

$$\begin{cases}\varphi_4(x)=\rho_0\int_0^\infty f_2(x+y)h_1(y)\,dy+\int_0^\infty \varphi_5(t)\gamma_2(x,t)\,dt,\\ \varphi_5(x)=\rho_0\int_0^\infty f_1(x+y)h_2(y)\,dy+\int_0^\infty \varphi_4(t)\gamma_1(x,t)\,dt.\end{cases} \tag{C.11}$$

Substitute the second equation of the system (C.11) into the first one. We have:

$$\varphi_4(x)=\rho_0\int_0^\infty f_2(x+y)h_1(y)dy+\rho_0\int_0^\infty \gamma_2(x,t)dt\int_0^\infty f_1(t+y)h_2(y)dy+$$
$$+\int_0^\infty \varphi_4(y)dy\int_0^\infty \gamma_2(x,t)\gamma_1(t,y)dt. \tag{C.12}$$

Introduce the function $k_2(x,y)=\int_0^\infty \gamma_2(x,t)\gamma_1(t,y)\,dt$

and the operator $(K_2\varphi)(x)=\int_0^\infty k_2(x,y)\varphi(y)\,dy$,

then the equation (C.12) can be rewritten in such a way:

$$\varphi_4(x)=(K_2\varphi_4)(x)+\rho_0\int_0^\infty f_2(x+y)h_1(y)\,dy$$
$$+\rho_0\int_0^\infty \gamma_2(x,t)\,dt\int_0^\infty f_1(t+y)h_2(y)\,dy, \tag{C.13}$$

(C.13) – Fredholm's equation of the second kind [11].
We shall find $\varphi_4(x)$.

$$(I-K_2)\varphi_4(x)=\rho_0\int_0^\infty f_2(x+y)h_1(y)\,dy+\rho_0\int_0^\infty \gamma_2(x,t)\,dt\int_0^\infty f_1(t+y)h_2(y)\,dy,$$

$$\varphi_2(x_2)=(I-k_2)^{-1}\left[\rho_0\int_0^\infty f_2(x_2+y)h_1(y)\,dy+\rho_0\int_0^\infty \gamma_2(x_2,t)\,dt\int_0^\infty f_1(t+y)h_2(y)\,dy\right].$$

Let us define the nth power of the operator K_2.

$$\left(K_2^2\varphi\right)(x)=\int_0^\infty k_2(x,y)\,dy\int_0^\infty k_2(y,t)\varphi(t)\,dt=\int_0^\infty \varphi(t)\,dt\int_0^\infty k_2(x,y)k_2(y,t)\,dy=$$

$$=\int_0^\infty \varphi(y)\,dy\int_0^\infty k_2(x,t)k_2(t,y)\,dt=\int_0^\infty k_2^{(2)}(x,y)\varphi(y)\,dy,\; k_2^{(2)}(x,y)=\int_0^\infty k_2(x,t)k_2(t,y)\,dt,$$

$$\left(K_2^n\varphi\right)(x)=\int_0^\infty k_2^{(n)}(x,y)\varphi(y)\,dy,\;\; k_2^{(n)}(x,y)=\int_0^\infty k_2^{(n-1)}(x,t)k_2(t,y)\,dt.$$

$k_2^{(n)}(x,y)$ – nth iteration of the kernel $k_2(x,y)$,

$$\left(\sum_{n=1}^\infty K_2^n\varphi\right)(x)=\int_0^\infty \pi_2(x,y)\varphi(y)\,dy,\;\; \pi_2(x,y)=\sum_{n=1}^\infty k_2^{(n)}(x,y).$$

Then we get:

$$\varphi_4(x)=\rho_0\psi_2(x)=\rho_0\left(\int_0^\infty f_2(x+y)h_1(y)\,dy+\int_0^\infty \gamma_2(x,t)\,dt\int_0^\infty f_1(t+y)h_2(y)\,dy+\right.$$

$$\left.+\int_0^\infty \pi_2(x,y)\,dy\int_0^\infty f_2(y+t)h_1(t)\,dt+\int_0^\infty \pi_2(x,y)\,dy\int_0^\infty \gamma_2(y,t)\,dt\int_0^\infty f_1(t+z)h_2(z)\,dz\right),$$

denote by $\psi_2(x)$ the function in parenthesis.

By substituting the first equation of the system (C.11) in the second one and making analogical transformations, we obtain:

$$\varphi_5(x)=\rho_0\psi_1(x)=\rho_0\left(\int_0^\infty f_1(x+y)h_2(y)\,dy+\int_0^\infty \gamma_1(x,t)\,dt\int_0^\infty f_2(t+y)h_1(y)\,dy+\right.$$

$$\left.+\int_0^\infty \pi_1(x,y)\,dy\int_0^\infty f_1(y+t)h_2(t)\,dt+\int_0^\infty \pi_1(x,y)\,dy\int_0^\infty \gamma_1(y,t)\,dt\int_0^\infty f_2(t+z)h_1(z)\,dz\right),$$

where

$$\pi_1(x,y)=\sum_{n=1}^\infty k_1^{(n)}(x,y),\;\; k_1(x,y)=\int_0^\infty \gamma_1(x,t)\gamma_2(t,y)\,dt,$$

$$k_1^{(n)}(x,y)=\int_0^\infty k_1^{(n-1)}(x,t)k_1(t,y)\,dt,\;\; k_1^{(1)}(x,y)=k_1(x,y).$$

Let us find the rest of the sought functions by expressing them in terms of $\psi_1(x)$ and $\psi_2(x)$. Applying (C.1), we get:

$$\varphi_1(x_1,x_2)=\rho_0\left(\int_0^\infty f_1(x_1+t)f_2(x_2+t)h_r(t)dt+\right.$$

$$\left.+\int_0^\infty \psi_1(x_1+t)f_2(x_2+t)h_r(t)\,dt+\int_0^\infty \psi_2(x_2+t)f_1(x_1+t)h_r(t)\,dt\right).$$

Applying (C.3) and (C.4), we get:

$$\varphi_2(x,z)=\rho_0\left(\int_0^\infty f_1(y)f_2(x+y)v_r(y,z)dy+\right.$$

$$\left.+\int_0^\infty \psi_1(t)f_2(x+t)v_r(t,z)\,dt+\int_0^\infty \psi_2(x+t)f_1(t)v_r(t,z)\,dt\right),$$

$$\varphi_3(x,z)=\rho_0\left(\int_0^\infty f_1(x+y)f_2(y)v_r(y,z)\,dy+\right.$$

$$\left.+\int_0^\infty \psi_2(t)f_1(x+t)v_r(t,z)\,dt+\int_0^\infty \psi_1(x+t)f_2(t)v_r(t,z)\,dt\right).$$

Using the seventh and eighth equations of the system (3.6), we get:

$$\varphi_6(z)=\int_0^\infty \varphi_3(t,t+z)\,dt=\rho_0\left(\int_0^\infty dt\int_0^\infty f_1(t+y)f_2(y)v_r(y,t+z)dy+\right.$$

$$\left.+\int_0^\infty dt\int_0^\infty \psi_1(t+y)f_2(y)v_r(y,t+z)\,dy+\int_0^\infty dt\int_0^\infty \psi_2(y)f_1(t+y)v_r(y,t+z)\,dy\right),$$

$$\varphi_7(z)=\int_0^\infty \varphi_2(t,t+z)\,dt=\rho_0\left(\int_0^\infty dt\int_0^\infty f_1(y)f_2(t+y)v_r(y,t+z)dy+\right.$$

$$\left.+\int_0^\infty dt\int_0^\infty \psi_1(y)f_2(t+y)v_r(y,t+z)\,dy+\int_0^\infty dt\int_0^\infty \psi_2(t+y)f_1(y)v_r(y,t+z)\,dy\right),$$

the value of the constant ρ_0 can be found by means of normalization equation.

Consequently, the solution of the system (3.6) is given by the formula (3.9).

Appendix D

The Solution of the System of Equation (3.34)

Let us prove the system (3.34) to have the solution given by formula (3.35). Introduce the notation: $\rho(3111)=\rho_0$.

By substituting the third equation of the system (3.34) in the fifth one, we get:

$$\rho(3\hat{0}\hat{1}0)=\int_0^\infty \rho(1011z)e^{-\lambda_2 z}dz=\rho_0\lambda_1\int_0^\infty e^{-\lambda_2 z}dz\int_0^\infty e^{-(\lambda_1+\lambda_2)t}r(z+t)dt=$$

$$=\rho_0\lambda_1\int_0^\infty e^{-\lambda_2 z}dz\int_z^\infty e^{-(\lambda_1+\lambda_2)(t-z)}r(t)dt=\rho_0\lambda_1\int_0^\infty e^{-(\lambda_1+\lambda_2)t}r(t)dt\int_0^t e^{\lambda_1 z}dz= \tag{D.1}$$

$$=\rho_0\left(\int_0^\infty e^{-\lambda_2 t}r(t)dt-\int_0^\infty e^{-(\lambda_1+\lambda_2)t}r(t)dt\right).$$

Similarly, substituting the sixth equation of the system in the fourth, we have:

$$\rho(3\hat{1}\hat{0}0)=\rho_0\left(\int_0^\infty e^{-\lambda_1 t}r(t)dt-\int_0^\infty e^{-(\lambda_1+\lambda_2)t}r(t)dt\right). \tag{D.2}$$

Note,

$$\rho(321\hat{2})=\rho(3\hat{0}\hat{1}0),\quad \rho(3\hat{1}22)=\rho(3\hat{1}\hat{0}0). \tag{D.3}$$

Applying the ninth and fourth equations of the system, we find $\rho(1001z)$:

$$\rho(1001z)=\lambda_1\int_0^\infty \rho(2101z+t)e^{-\lambda_1 t}dt=\rho_0\lambda_1\lambda_2\int_0^\infty e^{-\lambda_1 t}dt\int_0^\infty e^{-(\lambda_1+\lambda_2)y}r(z+t+y)dy=$$

$$=\rho_0\lambda_1\lambda_2\int_0^\infty e^{-\lambda_1 t}dt\int_t^\infty e^{-(\lambda_1+\lambda_2)(y-t)}r(z+y)dy=\rho_0\lambda_1\lambda_2\int_0^\infty e^{-(\lambda_1+\lambda_2)y}r(z+y)dy\int_0^y e^{\lambda_2 t}dt=$$

Semi-Markov Models. http://dx.doi.org/10.1016/B978-0-12-802212-2.00009-7
Copyright © 2015 Elsevier Inc. All rights reserved.

$$= \rho_0 \lambda_1 \left(\int_0^{\infty} e^{-\lambda_1 y} r(z+y)\,dy - \int_0^{\infty} e^{-(\lambda_1+\lambda_2)y} r(z+y)\,dy \right). \tag{D.4}$$

In the same way, applying the tenth and third equations of the system, we get:

$$\rho(2001z) = \rho_0 \lambda_2 \left(\int_0^{\infty} e^{-\lambda_2 y} r(z+y)\,dy - \int_0^{\infty} e^{-(\lambda_1+\lambda_2)y} r(z+y)\,dy \right). \tag{D.5}$$

Next,

$$\rho(3\hat{0}\hat{0}0) = \int_0^{\infty} \rho(1001z)\,dz + \int_0^{\infty} \rho(2001z)\,dz =$$

$$= \rho_0 \lambda_1 \int_0^{\infty} dz \left(\int_0^{\infty} e^{-\lambda_1 y} r(z+y)\,dy - \int_0^{\infty} e^{-(\lambda_1+\lambda_2)y} r(z+y)\,dy \right) +$$

$$+ \rho_0 \lambda_2 \int_0^{\infty} dz \left(\int_0^{\infty} e^{-\lambda_2 y} r(z+y)\,dy - \int_0^{\infty} e^{-(\lambda_1+\lambda_2)y} r(z+y)\,dy \right) = \tag{D.6}$$

$$= \rho_0 \left(\lambda_1 \int_0^{\infty} e^{-\lambda_1 y} \bar{R}(y)\,dy + \lambda_2 \int_0^{\infty} e^{-\lambda_2 y} \bar{R}(y)\,dy - (\lambda_1 + \lambda_2) \int_0^{\infty} e^{-(\lambda_1+\lambda_2)y} \bar{R}(y)\,dy \right) =$$

$$= \rho_0 \left(1 - \int_0^{\infty} e^{-\lambda_1 y} r(y)\,dy - \int_0^{\infty} e^{-\lambda_2 y} r(y)\,dy + \int_0^{\infty} e^{-(\lambda_1+\lambda_2)y} r(y)\,dy \right).$$

Besides,

$$\rho(3222) = \rho(3\hat{0}\hat{0}0), \quad \rho(1\hat{1}22x) = \rho(3222) \int_0^{\infty} g_2(t+x) g_1(t)\,dt,$$
$$\rho(22\hat{1}2x) = \rho(3222) \int_0^{\infty} g_1(t+x) g_2(t)\,dt. \tag{D.7}$$

The value of constant ρ_0 is defined by the normalization requirement.

We conclude from (D.1) to (D.7) that the solution of the system (3.34) is given by formula (3.35).

References

[1] Barlow R, Proschan F. Mathematical theory of reliability. New York: John Wiley and Sons; 1965.

[2] Barlow R, Belyaev YK, Bogatirev VA, et al. In: Ushakov IA, editor. Reliability of technical systems: manual. Moscow: Radio I Svyaz Press; 1985.

[3] Beichelt F, Franken P. Reliability and maintenance: Mathematical method. Moscow: Radio I Svyaz Press; 1988.

[4] Cherkesov GN. Reliability of hardware-software complexes. St. Petersburg: Peter; 2005.

[5] Cherkesov GN. Reliability of technical systems with time redundancy. Moscow: Sov. Radio; 1974.

[6] Chupirin VN, Nikiforov AD. Technical control in mechanical engineering: guide for design engineer. Moscow: Mashinostroenie; 1987.

[7] Cox D, Smith V. Renewal theory. Moscow: Sov. Radio; 1967.

[8] Feller W. An Introduction to probability theory and its applications. Moscow: Mir; 1978. 2.

[9] Gill F, Murray W, Wright M. Practical optimization. Moscow: Mir; 1985.

[10] Joa DD, Buzacott JA. Flexible manufacturing system with limited local buffers. Int J Prod Res 1991;24(1):107–17.

[11] Kantorovich LV, Akilov GP. Functional analysis. Moscow: Nauka; 1977.

[12] Kleinrock L. Queueing theory. Moscow: Mashinostroenie; 1979.

[13] Kopp VY, Obzherin YuE, Peschanskiy AI. Stochastic models of automized system with time reservation. Sevastopol: Publishing Office of Sevastopol State Technical University; 2000.

[14] Korlat AN, Kuznetsov VN, Novikov MM, Turbin AF. Semi-Markovian models of restorable and service systems. Kishinev: Shtiintsa; 1991.

[15] Koroluk VS. Stochastic system models. Kiev: Lybid; 1993.

[16] Koroluk VS, Turbin AF. Markovian restoration processes in the problems of system reliability. Kiev: Naukova Dumka; 1982.

[17] Koroluk VS, Turbin AF. Semi-Markov processes and their applications. Kiev: Naukova Dumka; 1976.

[18] Kovalenko IN, Kuznetsov NYu, Shurenkov VM. Random processes: manual. Kiev: Naukova Dumka; 1983.

[19] Kuznetsov VN, Turbin AF, Tsaturyan GZh. Semi-Markov models of renewal systems, vol. 81. USSR: Institute of Mathematics; 1981. p. 4.

[20] National standard 27.004-85. Operating systems. Introd. Moscow: Izd-vo standartov; 2002.

[21] Obzherin YE, Nikishenko AN. Model of multi-component systems with latent failures and component control. Optimization of production processes, vol. 13. Sevastopol: Izd-vo SevNTU; 2011. p. 95–101.

[22] Obzherin YE, Nikishenko AN. Semi-Markov model of multi-component system based on the characteristics of the individual components. Nondestructive testing and diagnostics. Minsk 2012;4:3–13.

[23] Reinschke K, Ushakov IA. In: Ushakov IA, editor. Assessment of reliability of systems using graphs. Moscow: Radio I Svyaz Press; 1988.

[24] Schriber TJ. Simulation Using GPSS. Moscow: Mashinostroenie; 1980.

[25] Sevastyanov BA. Ergodic theorem for Markov processes and its application to telephone systems with rejections. Probability Theory Appl 1957;2(1):106–16.

[26] Shurenkov VM. Ergodic Markovian processes. Moscow: Nauka; 1989.

[27] Yampolsky LC, Polishchuk MN. Optimization of operating processes in flexible operating systems. Kiev: Tekhnika; 1988.

Index

T

V

W

X

哈尔滨工业大学出版社刘培杰数学工作室已出版(即将出版)图书目录

书　　名	出版时间	定　价	编号
新编中学数学解题方法全书(高中版)上卷	2007—09	38.00	7
新编中学数学解题方法全书(高中版)中卷	2007—09	48.00	8
新编中学数学解题方法全书(高中版)下卷(一)	2007—09	42.00	17
新编中学数学解题方法全书(高中版)下卷(二)	2007—09	38.00	18
新编中学数学解题方法全书(高中版)下卷(三)	2010—06	58.00	73
新编中学数学解题方法全书(初中版)上卷	2008—01	28.00	29
新编中学数学解题方法全书(初中版)中卷	2010—07	38.00	75
新编中学数学解题方法全书(高考复习卷)	2010—01	48.00	67
新编中学数学解题方法全书(高考真题卷)	2010—01	38.00	62
新编中学数学解题方法全书(高考精华卷)	2011—03	68.00	118
新编平面解析几何解题方法全书(专题讲座卷)	2010—01	18.00	61
新编中学数学解题方法全书(自主招生卷)	2013—08	88.00	261
数学眼光透视	2008—01	38.00	24
数学思想领悟	2008—01	38.00	25
数学应用展观	2008—01	38.00	26
数学建模导引	2008—01	28.00	23
数学方法溯源	2008—01	38.00	27
数学史话览胜	2008—01	28.00	28
数学思维技术	2013—09	38.00	260
从毕达哥拉斯到怀尔斯	2007—10	48.00	9
从迪利克雷到维斯卡尔迪	2008—01	48.00	21
从哥德巴赫到陈景润	2008—05	98.00	35
从庞加莱到佩雷尔曼	2011—08	138.00	136
数学奥林匹克与数学文化(第一辑)	2006—05	48.00	4
数学奥林匹克与数学文化(第二辑)(竞赛卷)	2008—01	48.00	19
数学奥林匹克与数学文化(第二辑)(文化卷)	2008—07	58.00	36′
数学奥林匹克与数学文化(第三辑)(竞赛卷)	2010—01	48.00	59
数学奥林匹克与数学文化(第四辑)(竞赛卷)	2011—08	58.00	87
数学奥林匹克与数学文化(第五辑)	2015—06	98.00	370

哈尔滨工业大学出版社刘培杰数学工作室已出版(即将出版)图书目录

书　　名	出版时间	定　价	编号
世界著名平面几何经典著作钩沉——几何作图专题卷(上)	2009—06	48.00	49
世界著名平面几何经典著作钩沉——几何作图专题卷(下)	2011—01	88.00	80
世界著名平面几何经典著作钩沉(民国平面几何老课本)	2011—03	38.00	113
世界著名平面几何经典著作钩沉(建国初期平面三角老课本)	2015—08	38.00	507
世界著名解析几何经典著作钩沉——平面解析几何卷	2014—01	38.00	264
世界著名数论经典著作钩沉(算术卷)	2012—01	28.00	125
世界著名数学经典著作钩沉——立体几何卷	2011—02	28.00	88
世界著名三角学经典著作钩沉(平面三角卷Ⅰ)	2010—06	28.00	69
世界著名三角学经典著作钩沉(平面三角卷Ⅱ)	2011—01	38.00	78
世界著名初等数论经典著作钩沉(理论和实用算术卷)	2011—07	38.00	126

发展空间想象力	2010—01	38.00	57
走向国际数学奥林匹克的平面几何试题诠释(上、下)(第1版)	2007—01	68.00	11,12
走向国际数学奥林匹克的平面几何试题诠释(上、下)(第2版)	2010—02	98.00	63,64
平面几何证明方法全书	2007—08	35.00	1
平面几何证明方法全书习题解答(第1版)	2005—10	18.00	2
平面几何证明方法全书习题解答(第2版)	2006—12	18.00	10
平面几何天天练上卷・基础篇(直线型)	2013—01	58.00	208
平面几何天天练中卷・基础篇(涉及圆)	2013—01	28.00	234
平面几何天天练下卷・提高篇	2013—01	58.00	237
平面几何专题研究	2013—07	98.00	258
最新世界各国数学奥林匹克中的平面几何试题	2007—09	38.00	14
数学竞赛平面几何典型题及新颖解	2010—07	48.00	74
初等数学复习及研究(平面几何)	2008—09	58.00	38
初等数学复习及研究(立体几何)	2010—06	38.00	71
初等数学复习及研究(平面几何)习题解答	2009—01	48.00	42
几何学教程(平面几何卷)	2011—03	68.00	90
几何学教程(立体几何卷)	2011—07	68.00	130
几何变换与几何证题	2010—06	88.00	70
计算方法与几何证题	2011—06	28.00	129
立体几何技巧与方法	2014—04	88.00	293
几何瑰宝——平面几何500名题暨1000条定理(上、下)	2010—07	138.00	76,77
三角形的解法与应用	2012—07	18.00	183
近代的三角形几何学	2012—07	48.00	184
一般折线几何学	2015—08	48.00	203
三角形的五心	2009—06	28.00	51
三角形的六心及其应用	2015—10	68.00	542
三角形趣谈	2012—08	28.00	212
解三角形	2014—01	28.00	265
三角学专门教程	2014—09	28.00	387

哈尔滨工业大学出版社刘培杰数学工作室已出版(即将出版)图书目录

书　名	出版时间	定　价	编号
距离几何分析导引	2015－02	68.00	446
圆锥曲线习题集(上册)	2013－06	68.00	255
圆锥曲线习题集(中册)	2015－01	78.00	434
圆锥曲线习题集(下册)	即将出版		
近代欧氏几何学	2012－03	48.00	162
罗巴切夫斯基几何学及几何基础概要	2012－07	28.00	188
罗巴切夫斯基几何学初步	2015－06	28.00	474
用三角、解析几何、复数、向量计算解数学竞赛几何题	2015－03	48.00	455
美国中学几何教程	2015－04	88.00	458
三线坐标与三角形特征点	2015－04	98.00	460
平面解析几何方法与研究(第1卷)	2015－05	18.00	471
平面解析几何方法与研究(第2卷)	2015－06	18.00	472
平面解析几何方法与研究(第3卷)	2015－07	18.00	473
解析几何研究	2015－01	38.00	425
解析几何学教程.上	2016－01	38.00	574
解析几何学教程.下	2016－01	38.00	575
几何学基础	2016－01	58.00	581
初等几何研究	2015－02	58.00	444
俄罗斯平面几何问题集	2009－08	88.00	55
俄罗斯立体几何问题集	2014－03	58.00	283
俄罗斯几何大师——沙雷金论数学及其他	2014－01	48.00	271
来自俄罗斯的5000道几何习题及解答	2011－03	58.00	89
俄罗斯初等数学问题集	2012－05	38.00	177
俄罗斯函数问题集	2011－03	38.00	103
俄罗斯组合分析问题集	2011－01	48.00	79
俄罗斯初等数学万题选——三角卷	2012－11	38.00	222
俄罗斯初等数学万题选——代数卷	2013－08	68.00	225
俄罗斯初等数学万题选——几何卷	2014－01	68.00	226
463个俄罗斯几何老问题	2012－01	28.00	152
超越吉米多维奇.数列的极限	2009－11	48.00	58
超越普里瓦洛夫.留数卷	2015－01	28.00	437
超越普里瓦洛夫.无穷乘积与它对解析函数的应用卷	2015－05	28.00	477
超越普里瓦洛夫.积分卷	2015－06	18.00	481
超越普里瓦洛夫.基础知识卷	2015－06	28.00	482
超越普里瓦洛夫.数项级数卷	2015－07	38.00	489
初等数论难题集(第一卷)	2009－05	68.00	44
初等数论难题集(第二卷)(上、下)	2011－02	128.00	82,83
数论概貌	2011－03	18.00	93
代数数论(第二版)	2013－08	58.00	94
代数多项式	2014－06	38.00	289
初等数论的知识与问题	2011－02	28.00	95
超越数论基础	2011－03	28.00	96
数论初等教程	2011－03	28.00	97
数论基础	2011－03	18.00	98
数论基础与维诺格拉多夫	2014－03	18.00	292

哈尔滨工业大学出版社刘培杰数学工作室已出版(即将出版)图书目录

书　　名	出版时间	定　价	编号
解析数论基础	2012－08	28.00	216
解析数论基础(第二版)	2014－01	48.00	287
解析数论问题集(第二版)	2014－05	88.00	343
数论入门	2011－03	38.00	99
代数数论入门	2015－03	38.00	448
数论开篇	2012－07	28.00	194
解析数论引论	2011－03	48.00	100
Barban Davenport Halberstam 均值和	2009－01	40.00	33
基础数论	2011－03	28.00	101
初等数论 100 例	2011－05	18.00	122
初等数论经典例题	2012－07	18.00	204
最新世界各国数学奥林匹克中的初等数论试题(上、下)	2012－01	138.00	144,145
初等数论(Ⅰ)	2012－01	18.00	156
初等数论(Ⅱ)	2012－01	18.00	157
初等数论(Ⅲ)	2012－01	28.00	158
平面几何与数论中未解决的新老问题	2013－01	68.00	229
代数数论简史	2014－11	28.00	408
代数数论	2015－09	88.00	532
数论导引提要及习题解答	2016－01	48.00	559
谈谈素数	2011－03	18.00	91
平方和	2011－03	18.00	92
复变函数引论	2013－10	68.00	269
伸缩变换与抛物旋转	2015－01	38.00	449
无穷分析引论(上)	2013－04	88.00	247
无穷分析引论(下)	2013－04	98.00	245
数学分析	2014－04	28.00	338
数学分析中的一个新方法及其应用	2013－01	38.00	231
数学分析例选:通过范例学技巧	2013－01	88.00	243
高等代数例选:通过范例学技巧	2015－06	88.00	475
三角级数论(上册)(陈建功)	2013－01	38.00	232
三角级数论(下册)(陈建功)	2013－01	48.00	233
三角级数论(哈代)	2013－06	48.00	254
三角级数	2015－07	28.00	263
超越数	2011－03	18.00	109
三角和方法	2011－03	18.00	112
整数论	2011－05	38.00	120
从整数谈起	2015－10	28.00	538
随机过程(Ⅰ)	2014－01	78.00	224
随机过程(Ⅱ)	2014－01	68.00	235
算术探索	2011－12	158.00	148
组合数学	2012－04	28.00	178
组合数学浅谈	2012－03	28.00	159
丢番图方程引论	2012－03	48.00	172
拉普拉斯变换及其应用	2015－02	38.00	447
高等代数.上	2016－01	38.00	548
高等代数.下	2016－01	38.00	549

哈尔滨工业大学出版社刘培杰数学工作室已出版(即将出版)图书目录

书　名	出版时间	定　价	编号
高等代数教程	2016－01	58.00	579
数学解析教程.上卷.1	2016－01	58.00	546
数学解析教程.上卷.2	2016－01	38.00	553
函数构造论.上	2016－01	38.00	554
函数构造论.下	即将出版		555
数与多项式	2016－01	38.00	558
概周期函数	2016－01	48.00	572
变叙的项的极限分布律	2016－01	18.00	573
整函数	2012－08	18.00	161
近代拓扑学研究	2013－04	38.00	239
多项式和无理数	2008－01	68.00	22
模糊数据统计学	2008－03	48.00	31
模糊分析学与特殊泛函空间	2013－01	68.00	241
谈谈不定方程	2011－05	28.00	119
常微分方程	2016－01	58.00	586
平稳随机函数导论	2016－03	48.00	587
量子力学原理·上	2016－01	38.00	588
受控理论与解析不等式	2012－05	78.00	165
解析不等式新论	2009－06	68.00	48
建立不等式的方法	2011－03	98.00	104
数学奥林匹克不等式研究	2009－08	68.00	56
不等式研究(第二辑)	2012－02	68.00	153
不等式的秘密(第一卷)	2012－02	28.00	154
不等式的秘密(第一卷)(第2版)	2014－02	38.00	286
不等式的秘密(第二卷)	2014－01	38.00	268
初等不等式的证明方法	2010－06	38.00	123
初等不等式的证明方法(第二版)	2014－11	38.00	407
不等式·理论·方法(基础卷)	2015－07	38.00	496
不等式·理论·方法(经典不等式卷)	2015－07	38.00	497
不等式·理论·方法(特殊类型不等式卷)	2015－07	48.00	498
同余理论	2012－05	38.00	163
[x]与{x}	2015－04	48.00	476
极值与最值.上卷	2015－06	28.00	486
极值与最值.中卷	2015－06	38.00	487
极值与最值.下卷	2015－06	28.00	488
整数的性质	2012－11	38.00	192
多项式理论	2015－10	88.00	541
历届美国中学生数学竞赛试题及解答(第一卷)1950－1954	2014－07	18.00	277
历届美国中学生数学竞赛试题及解答(第二卷)1955－1959	2014－04	18.00	278
历届美国中学生数学竞赛试题及解答(第三卷)1960－1964	2014－06	18.00	279
历届美国中学生数学竞赛试题及解答(第四卷)1965－1969	2014－04	28.00	280
历届美国中学生数学竞赛试题及解答(第五卷)1970－1972	2014－06	18.00	281
历届美国中学生数学竞赛试题及解答(第七卷)1981－1986	2015－01	18.00	424

哈尔滨工业大学出版社刘培杰数学工作室
已出版(即将出版)图书目录

书　　名	出版时间	定　价	编号
历届 IMO 试题集(1959—2005)	2006—05	58.00	5
历届 CMO 试题集	2008—09	28.00	40
历届中国数学奥林匹克试题集	2014—10	38.00	394
历届加拿大数学奥林匹克试题集	2012—08	38.00	215
历届美国数学奥林匹克试题集:多解推广加强	2012—08	38.00	209
历届波兰数学竞赛试题集. 第 1 卷,1949～1963	2015—03	18.00	453
历届波兰数学竞赛试题集. 第 2 卷,1964～1976	2015—03	18.00	454
历届巴尔干数学奥林匹克试题集	2015—05	38.00	466
保加利亚数学奥林匹克	2014—10	38.00	393
圣彼得堡数学奥林匹克试题集	2015—01	38.00	429
历届国际大学生数学竞赛试题集(1994—2010)	2012—01	28.00	143
全国大学生数学夏令营数学竞赛试题及解答	2007—03	28.00	15
全国大学生数学竞赛辅导教程	2012—07	28.00	189
全国大学生数学竞赛复习全书	2014—04	48.00	340
历届美国大学生数学竞赛试题集	2009—03	88.00	43
前苏联大学生数学奥林匹克竞赛题解(上编)	2012—04	28.00	169
前苏联大学生数学奥林匹克竞赛题解(下编)	2012—04	38.00	170
历届美国数学邀请赛试题集	2014—01	48.00	270
全国高中数学竞赛试题及解答. 第 1 卷	2014—07	38.00	331
大学生数学竞赛讲义	2014—09	28.00	371
亚太地区数学奥林匹克竞赛题	2015—07	18.00	492

高考数学临门一脚(含密押三套卷)(理科版)	2015—01	24.80	421
高考数学临门一脚(含密押三套卷)(文科版)	2015—01	24.80	422
新课标高考数学题型全归纳(文科版)	2015—05	72.00	467
新课标高考数学题型全归纳(理科版)	2015—05	82.00	468
王连笑教你怎样学数学:高考选择题解题策略与客观题实用训练	2014—01	48.00	262
王连笑教你怎样学数学:高考数学高层次讲座	2015—02	48.00	432
高考数学的理论与实践	2009—08	38.00	53
高考数学核心题型解题方法与技巧	2010—01	28.00	86
高考思维新平台	2014—03	38.00	259
30 分钟拿下高考数学选择题、填空题(第二版)	2012—01	28.00	146
高考数学压轴题解题诀窍(上)	2012—02	78.00	166
高考数学压轴题解题诀窍(下)	2012—03	28.00	167
北京市五区文科数学三年高考模拟题详解:2013～2015	2015—08	48.00	500
北京市五区理科数学三年高考模拟题详解:2013～2015	2015—09	68.00	505
向量法巧解数学高考题	2009—08	28.00	54
高考数学万能解题法	2015—09	28.00	534
高考物理万能解题法	2015—09	28.00	537
高考化学万能解题法	2015—11	25.00	557
我一定要赚分:高中物理	2016—01	38.00	580
数学高考参考	2016—01	78.00	589
2011～2015 年全国及各省市高考数学文科精品试题审题要津与解法研究	2015—10	68.00	539
2011～2015 年全国及各省市高考数学理科精品试题审题要津与解法研究	2015—10	88.00	540
最新全国及各省市高考数学试卷解法研究及点拨评析	2009—02	38.00	41
2011 年全国及各省市高考数学试题审题要津与解法研究	2011—10	48.00	139

哈尔滨工业大学出版社刘培杰数学工作室
已出版(即将出版)图书目录

书　　名	出版时间	定　价	编号
2013 年全国及各省市高考数学试题解析与点评	2014－01	48.00	282
全国及各省市高考数学试题审题要津与解法研究	2015－02	48.00	450
新课标高考数学——五年试题分章详解(2007～2011)(上、下)	2011－10	78.00	140,141
全国中考数学压轴题审题要津与解法研究	2013－04	78.00	248
新编全国及各省市中考数学压轴题审题要津与解法研究	2014－05	58.00	342
全国及各省市 5 年中考数学压轴题审题要津与解法研究	2015－04	58.00	462
中考数学专题总复习	2007－04	28.00	6
中考数学较难题、难题常考题型解题方法与技巧.上	2016－01	48.00	584
中考数学较难题、难题常考题型解题方法与技巧.下	2016－01	58.00	585
数学奥林匹克在中国	2014－06	98.00	344
数学奥林匹克问题集	2014－01	38.00	267
数学奥林匹克不等式散论	2010－06	38.00	124
数学奥林匹克不等式欣赏	2011－09	38.00	138
数学奥林匹克超级题库(初中卷上)	2010－01	58.00	66
数学奥林匹克不等式证明方法和技巧(上、下)	2011－08	158.00	134,135
新编 640 个世界著名数学智力趣题	2014－01	88.00	242
500 个最新世界著名数学智力趣题	2008－06	48.00	3
400 个最新世界著名数学最值问题	2008－09	48.00	36
500 个世界著名数学征解问题	2009－06	48.00	52
400 个中国最佳初等数学征解老问题	2010－01	48.00	60
500 个俄罗斯数学经典老题	2011－01	28.00	81
1000 个国外中学物理好题	2012－04	48.00	174
300 个日本高考数学题	2012－05	38.00	142
500 个前苏联早期高考数学试题及解答	2012－05	28.00	185
546 个早期俄罗斯大学生数学竞赛题	2014－03	38.00	285
548 个来自美苏的数学好问题	2014－11	28.00	396
20 所苏联著名大学早期入学试题	2015－02	18.00	452
161 道德国工科大学生必做的微分方程习题	2015－05	28.00	469
500 个德国工科大学生必做的高数习题	2015－06	28.00	478
德国讲义日本考题.微积分卷	2015－04	48.00	456
德国讲义日本考题.微分方程卷	2015－04	38.00	457
几何变换(Ⅰ)	2014－07	28.00	353
几何变换(Ⅱ)	2015－06	28.00	354
几何变换(Ⅲ)	2015－01	38.00	355
几何变换(Ⅳ)	2015－12	38.00	356
中国初等数学研究　2009 卷(第 1 辑)	2009－05	20.00	45
中国初等数学研究　2010 卷(第 2 辑)	2010－05	30.00	68
中国初等数学研究　2011 卷(第 3 辑)	2011－07	60.00	127
中国初等数学研究　2012 卷(第 4 辑)	2012－07	48.00	190
中国初等数学研究　2014 卷(第 5 辑)	2014－02	48.00	288
中国初等数学研究　2015 卷(第 6 辑)	2015－06	68.00	493
博弈论精粹	2008－03	58.00	30
博弈论精粹.第二版(精装)	2015－01	88.00	461
数学 我爱你	2008－01	28.00	20

哈尔滨工业大学出版社刘培杰数学工作室已出版(即将出版)图书目录

书　名	出版时间	定　价	编号
精神的圣徒　别样的人生——60位中国数学家成长的历程	2008－09	48.00	39
数学史概论	2009－06	78.00	50
数学史概论(精装)	2013－03	158.00	272
数学史选讲	2016－01	48.00	544
斐波那契数列	2010－02	28.00	65
数学拼盘和斐波那契魔方	2010－07	38.00	72
斐波那契数列欣赏	2011－01	28.00	160
数学的创造	2011－02	48.00	85
数海拾贝	2016－01	48.00	590
数学中的美	2011－02	38.00	84
数论中的美学	2014－12	38.00	351
数学王者　科学巨人——高斯	2015－01	28.00	428
振兴祖国数学的圆梦之旅:中国初等数学研究史话	2015－06	78.00	490
二十世纪中国数学史料研究	2015－10	48.00	536
数字谜、数阵图与棋盘覆盖	2016－01	58.00	298
时间的形状	2016－01	38.00	556
数学解题——靠数学思想给力(上)	2011－07	38.00	131
数学解题——靠数学思想给力(中)	2011－07	48.00	132
数学解题——靠数学思想给力(下)	2011－07	38.00	133
我怎样解题	2013－01	48.00	227
数学解题中的物理方法	2011－06	28.00	114
数学解题的特殊方法	2011－06	48.00	115
中学数学计算技巧	2012－01	48.00	116
中学数学证明方法	2012－01	58.00	117
数学趣题巧解	2012－03	28.00	128
高中数学教学通鉴	2015－05	58.00	479
和高中生漫谈:数学与哲学的故事	2014－08	28.00	369
自主招生考试中的参数方程问题	2015－01	28.00	435
自主招生考试中的极坐标问题	2015－04	28.00	463
近年全国重点大学自主招生数学试题全解及研究.华约卷	2015－02	38.00	441
近年全国重点大学自主招生数学试题全解及研究.北约卷	即将出版		
自主招生数学解证宝典	2015－09	48.00	535
格点和面积	2012－07	18.00	191
射影几何趣谈	2012－04	28.00	175
斯潘纳尔引理——从一道加拿大数学奥林匹克试题谈起	2014－01	28.00	228
李普希兹条件——从几道近年高考数学试题谈起	2012－10	18.00	221
拉格朗日中值定理——从一道北京高考试题的解法谈起	2015－10	18.00	197
闵科夫斯基定理——从一道清华大学自主招生试题谈起	2014－01	28.00	198
哈尔测度——从一道冬令营试题的背景谈起	2012－08	28.00	202
切比雪夫逼近问题——从一道中国台北数学奥林匹克试题谈起	2013－04	38.00	238
伯恩斯坦多项式与贝齐尔曲面——从一道全国高中数学联赛试题谈起	2013－03	38.00	236
卡塔兰猜想——从一道普特南竞赛试题谈起	2013－06	18.00	256

哈尔滨工业大学出版社刘培杰数学工作室已出版(即将出版)图书目录

书　　名	出版时间	定　价	编号
麦卡锡函数和阿克曼函数——从一道前南斯拉夫数学奥林匹克试题谈起	2012－08	18.00	201
贝蒂定理与拉姆贝克莫斯尔定理——从一个拣石子游戏谈起	2012－08	18.00	217
皮亚诺曲线和豪斯道夫分球定理——从无限集谈起	2012－08	18.00	211
平面凸图形与凸多面体	2012－10	28.00	218
斯坦因豪斯问题——从一道二十五省市自治区中学数学竞赛试题谈起	2012－07	18.00	196
纽结理论中的亚历山大多项式与琼斯多项式——从一道北京市高一数学竞赛试题谈起	2012－07	28.00	195
原则与策略——从波利亚"解题表"谈起	2013－04	38.00	244
转化与化归——从三大尺规作图不能问题谈起	2012－08	28.00	214
代数几何中的贝祖定理(第一版)——从一道 IMO 试题的解法谈起	2013－08	18.00	193
成功连贯理论与约当块理论——从一道比利时数学竞赛试题谈起	2012－04	18.00	180
磨光变换与范·德·瓦尔登猜想——从一道环球城市竞赛试题谈起	即将出版		
素数判定与大数分解	2014－08	18.00	199
置换多项式及其应用	2012－10	18.00	220
椭圆函数与模函数——从一道美国加州大学洛杉矶分校(UCLA)博士资格考题谈起	2012－10	28.00	219
差分方程的拉格朗日方法——从一道 2011 年全国高考理科试题的解法谈起	2012－08	28.00	200
力学在几何中的一些应用	2013－01	38.00	240
高斯散度定理、斯托克斯定理和平面格林定理——从一道国际大学生数学竞赛试题谈起	即将出版		
康托洛维奇不等式——从一道全国高中联赛试题谈起	2013－03	28.00	337
西格尔引理——从一道第 18 届 IMO 试题的解法谈起	即将出版		
罗斯定理——从一道前苏联数学竞赛试题谈起	即将出版		
拉克斯定理和阿廷定理——从一道 IMO 试题的解法谈起	2014－01	58.00	246
毕卡大定理——从一道美国大学数学竞赛试题谈起	2014－07	18.00	350
贝齐尔曲线——从一道全国高中联赛试题谈起	即将出版		
拉格朗日乘子定理——从一道 2005 年全国高中联赛试题的高等数学解法谈起	2015－05	28.00	480
雅可比定理——从一道日本数学奥林匹克试题谈起	2013－04	48.00	249
李天岩－约克定理——从一道波兰数学竞赛试题谈起	2014－06	28.00	349
整系数多项式因式分解的一般方法——从克朗耐克算法谈起	即将出版		
布劳维不动点定理——从一道前苏联数学奥林匹克试题谈起	2014－01	38.00	273
压缩不动点定理——从一道高考数学试题的解法谈起	即将出版		
伯恩赛德定理——从一道英国数学奥林匹克试题谈起	即将出版		
布查特－莫斯特定理——从一道上海市初中竞赛试题谈起	即将出版		
数论中的同余数问题——从一道普特南竞赛试题谈起	即将出版		
范·德蒙行列式——从一道美国数学奥林匹克试题谈起	即将出版		
中国剩余定理:总数法构建中国历史年表	2015－01	28.00	430
牛顿程序与方程求根——从一道全国高考试题解法谈起	即将出版		
库默尔定理——从一道 IMO 预选试题谈起	即将出版		

哈尔滨工业大学出版社刘培杰数学工作室
已出版(即将出版)图书目录

书 名	出版时间	定 价	编号
卢丁定理——从一道冬令营试题的解法谈起	即将出版		
沃斯滕霍姆定理——从一道 IMO 预选试题谈起	即将出版		
卡尔松不等式——从一道莫斯科数学奥林匹克试题谈起	即将出版		
信息论中的香农熵——从一道近年高考压轴题谈起	即将出版		
约当不等式——从一道希望杯竞赛试题谈起	即将出版		
拉比诺维奇定理	即将出版		
刘维尔定理——从一道《美国数学月刊》征解问题的解法谈起	即将出版		
卡塔兰恒等式与级数求和——从一道 IMO 试题的解法谈起	即将出版		
勒让德猜想与素数分布——从一道爱尔兰竞赛试题谈起	即将出版		
天平称重与信息论——从一道基辅市数学奥林匹克试题谈起	即将出版		
哈密尔顿一凯莱定理:从一道高中数学联赛试题的解法谈起	2014－09	18.00	376
艾思特曼定理——从一道 CMO 试题的解法谈起	即将出版		
一个爱尔特希问题——从一道西德数学奥林匹克试题谈起	即将出版		
有限群中的爱丁格尔问题——从一道北京市初中二年级数学竞赛试题谈起	即将出版		
贝克码与编码理论——从一道全国高中联赛试题谈起	即将出版		
帕斯卡三角形	2014－03	18.00	294
蒲丰投针问题——从 2009 年清华大学的一道自主招生试题谈起	2014－01	38.00	295
斯图姆定理——从一道"华约"自主招生试题的解法谈起	2014－01	18.00	296
许瓦兹引理——从一道加利福尼亚大学伯克利分校数学系博士生试题谈起	2014－08	18.00	297
拉姆塞定理——从王诗宬院士的一个问题谈起	2014－01		299
坐标法	2013－12	28.00	332
数论三角形	2014－04	38.00	341
毕克定理	2014－07	18.00	352
数林掠影	2014－09	48.00	389
我们周围的概率	2014－10	38.00	390
凸函数最值定理:从一道华约自主招生题的解法谈起	2014－10	28.00	391
易学与数学奥林匹克	2014－10	38.00	392
生物数学趣谈	2015－01	18.00	409
反演	2015－01		420
因式分解与圆锥曲线	2015－01	18.00	426
轨迹	2015－01	28.00	427
面积原理:从常庚哲命的一道 CMO 试题的积分解法谈起	2015－01	48.00	431
形形色色的不动点定理:从一道 28 届 IMO 试题谈起	2015－01	38.00	439
柯西函数方程:从一道上海交大自主招生的试题谈起	2015－02	28.00	440
三角恒等式	2015－02	28.00	442
无理性判定:从一道 2014 年"北约"自主招生试题谈起	2015－01	38.00	443
数学归纳法	2015－03	18.00	451
极端原理与解题	2015－04	28.00	464
法雷级数	2014－08	18.00	367
摆线族	2015－01	38.00	438
函数方程及其解法	2015－05	38.00	470
含参数的方程和不等式	2012－09	28.00	213
希尔伯特第十问题	2016－01	38.00	543
无穷小量的求和	2016－01	28.00	545
切比雪夫多项式:从一道清华大学金秋营试题谈起	2016－01	38.00	583

哈尔滨工业大学出版社刘培杰数学工作室
已出版(即将出版)图书目录

书　　名	出版时间	定　价	编号
中等数学英语阅读文选	2006－12	38.00	13
统计学专业英语	2007－03	28.00	16
统计学专业英语(第二版)	2012－07	48.00	176
统计学专业英语(第三版)	2015－04	68.00	465
幻方和魔方(第一卷)	2012－05	68.00	173
尘封的经典——初等数学经典文献选读(第一卷)	2012－07	48.00	205
尘封的经典——初等数学经典文献选读(第二卷)	2012－07	38.00	206
代换分析:英文	2015－07	38.00	499

实变函数论	2012－06	78.00	181
复变函数论	2015－08	38.00	504
非光滑优化及其变分分析	2014－01	48.00	230
疏散的马尔科夫链	2014－01	58.00	266
马尔科夫过程论基础	2015－01	28.00	433
初等微分拓扑学	2012－07	18.00	182
方程式论	2011－03	38.00	105
初级方程式论	2011－03	28.00	106
Galois 理论	2011－03	18.00	107
古典数学难题与伽罗瓦理论	2012－11	58.00	223
伽罗华与群论	2014－01	28.00	290
代数方程的根式解及伽罗瓦理论	2011－03	28.00	108
代数方程的根式解及伽罗瓦理论(第二版)	2015－01	28.00	423
线性偏微分方程讲义	2011－03	18.00	110
几类微分方程数值方法的研究	2015－05	38.00	485
N 体问题的周期解	2011－03	28.00	111
代数方程式论	2011－05	18.00	121
动力系统的不变量与函数方程	2011－07	48.00	137
基于短语评价的翻译知识获取	2012－02	48.00	168
应用随机过程	2012－04	48.00	187
概率论导引	2012－04	18.00	179
矩阵论(上)	2013－06	58.00	250
矩阵论(下)	2013－06	48.00	251
对称锥互补问题的内点法:理论分析与算法实现	2014－08	68.00	368
抽象代数:方法导引	2013－06	38.00	257
集论	2016－01	48.00	576
多项式理论研究综述	2016－01	38.00	577
函数论	2014－11	78.00	395
反问题的计算方法及应用	2011－11	28.00	147
初等数学研究(Ⅰ)	2008－09	68.00	37
初等数学研究(Ⅱ)(上、下)	2009－05	118.00	46,47
数阵及其应用	2012－02	28.00	164
绝对值方程—折边与组合图形的解析研究	2012－07	48.00	186
代数函数论(上)	2015－07	38.00	494
代数函数论(下)	2015－07	38.00	495
偏微分方程论:法文	2015－10	48.00	533

闵嗣鹤文集	2011－03	98.00	102
吴从炘数学活动三十年(1951～1980)	2010－07	99.00	32
吴从炘数学活动又三十年(1981～2010)	2015－07	98.00	491

哈尔滨工业大学出版社刘培杰数学工作室
已出版(即将出版)图书目录

书名	出版时间	定价	编号
趣味初等方程妙题集锦	2014－09	48.00	388
趣味初等数论选美与欣赏	2015－02	48.00	445
耕读笔记(上卷):一位农民数学爱好者的初数探索	2015－04	28.00	459
耕读笔记(中卷):一位农民数学爱好者的初数探索	2015－05	28.00	483
耕读笔记(下卷):一位农民数学爱好者的初数探索	2015－05	28.00	484
几何不等式研究与欣赏.上卷	2016－01	88.00	547
几何不等式研究与欣赏.下卷	2016－01	48.00	552
初等数列研究与欣赏·上	2016－01	48.00	570
初等数列研究与欣赏·下	2016－01	48.00	571
数贝偶拾——高考数学题研究	2014－04	28.00	274
数贝偶拾——初等数学研究	2014－04	38.00	275
数贝偶拾——奥数题研究	2014－04	48.00	276
集合、函数与方程	2014－01	28.00	300
数列与不等式	2014－01	38.00	301
三角与平面向量	2014－01	28.00	302
平面解析几何	2014－01	38.00	303
立体几何与组合	2014－01	28.00	304
极限与导数、数学归纳法	2014－01	38.00	305
趣味数学	2014－03	28.00	306
教材教法	2014－04	68.00	307
自主招生	2014－05	58.00	308
高考压轴题(上)	2015－01	48.00	309
高考压轴题(下)	2014－10	68.00	310
从费马到怀尔斯——费马大定理的历史	2013－10	198.00	Ⅰ
从庞加莱到佩雷尔曼——庞加莱猜想的历史	2013－10	298.00	Ⅱ
从切比雪夫到爱尔特希(上)——素数定理的初等证明	2013－07	48.00	Ⅲ
从切比雪夫到爱尔特希(下)——素数定理100年	2012－12	98.00	Ⅲ
从高斯到盖尔方特——二次域的高斯猜想	2013－10	198.00	Ⅳ
从库默尔到朗兰兹——朗兰兹猜想的历史	2014－01	98.00	Ⅴ
从比勃巴赫到德布朗斯——比勃巴赫猜想的历史	2014－02	298.00	Ⅵ
从麦比乌斯到陈省身——麦比乌斯变换与麦比乌斯带	2014－02	298.00	Ⅶ
从布尔到豪斯道夫——布尔方程与格论漫谈	2013－10	198.00	Ⅷ
从开普勒到阿诺德——三体问题的历史	2014－05	298.00	Ⅸ
从华林到华罗庚——华林问题的历史	2013－10	298.00	Ⅹ
第19~23届"希望杯"全国数学邀请赛试题审题要津详细评注(初一版)	2014－03	28.00	333
第19~23届"希望杯"全国数学邀请赛试题审题要津详细评注(初二、初三版)	2014－03	38.00	334
第19~23届"希望杯"全国数学邀请赛试题审题要津详细评注(高一版)	2014－03	28.00	335
第19~23届"希望杯"全国数学邀请赛试题审题要津详细评注(高二版)	2014－03	38.00	336
第19~25届"希望杯"全国数学邀请赛试题审题要津详细评注(初一版)	2015－01	38.00	416
第19~25届"希望杯"全国数学邀请赛试题审题要津详细评注(初二、初三版)	2015－01	58.00	417
第19~25届"希望杯"全国数学邀请赛试题审题要津详细评注(高一版)	2015－01	48.00	418
第19~25届"希望杯"全国数学邀请赛试题审题要津详细评注(高二版)	2015－01	48.00	419

哈尔滨工业大学出版社刘培杰数学工作室已出版(即将出版)图书目录

书　名	出版时间	定　价	编号
吴振奎高等数学解题真经(概率统计卷)	2012－01	38.00	149
吴振奎高等数学解题真经(微积分卷)	2012－01	68.00	150
吴振奎高等数学解题真经(线性代数卷)	2012－01	58.00	151
钱昌本教你快乐学数学(上)	2011－12	48.00	155
钱昌本教你快乐学数学(下)	2012－03	58.00	171
高等数学解题全攻略(上卷)	2013－06	58.00	252
高等数学解题全攻略(下卷)	2013－06	58.00	253
高等数学复习纲要	2014－01	18.00	384
三角函数	2014－01	38.00	311
不等式	2014－01	38.00	312
数列	2014－01	38.00	313
方程	2014－01	28.00	314
排列和组合	2014－01	28.00	315
极限与导数	2014－01	28.00	316
向量	2014－09	38.00	317
复数及其应用	2014－08	28.00	318
函数	2014－01	38.00	319
集合	即将出版		320
直线与平面	2014－01	28.00	321
立体几何	2014－04	28.00	322
解三角形	即将出版		323
直线与圆	2014－01	28.00	324
圆锥曲线	2014－01	38.00	325
解题通法(一)	2014－07	38.00	326
解题通法(二)	2014－07	38.00	327
解题通法(三)	2014－05	38.00	328
概率与统计	2014－01	28.00	329
信息迁移与算法	即将出版		330
物理奥林匹克竞赛大题典——力学卷	2014－11	48.00	405
物理奥林匹克竞赛大题典——热学卷	2014－04	28.00	339
物理奥林匹克竞赛大题典——电磁学卷	2015－07	48.00	406
物理奥林匹克竞赛大题典——光学与近代物理卷	2014－06	28.00	345
历届中国东南地区数学奥林匹克试题集(2004～2012)	2014－06	18.00	346
历届中国西部地区数学奥林匹克试题集(2001～2012)	2014－07	18.00	347
历届中国女子数学奥林匹克试题集(2002～2012)	2014－08	18.00	348
美国高中数学竞赛五十讲. 第 1 卷(英文)	2014－08	28.00	357
美国高中数学竞赛五十讲. 第 2 卷(英文)	2014－08	28.00	358
美国高中数学竞赛五十讲. 第 3 卷(英文)	2014－09	28.00	359
美国高中数学竞赛五十讲. 第 4 卷(英文)	2014－09	28.00	360
美国高中数学竞赛五十讲. 第 5 卷(英文)	2014－10	28.00	361
美国高中数学竞赛五十讲. 第 6 卷(英文)	2014－11	28.00	362
美国高中数学竞赛五十讲. 第 7 卷(英文)	2014－12	28.00	363
美国高中数学竞赛五十讲. 第 8 卷(英文)	2015－01	28.00	364
美国高中数学竞赛五十讲. 第 9 卷(英文)	2015－01	28.00	365
美国高中数学竞赛五十讲. 第 10 卷(英文)	2015－02	38.00	366

哈尔滨工业大学出版社刘培杰数学工作室
已出版(即将出版)图书目录

书　　名	出版时间	定　价	编号
IMO 50 年. 第 1 卷(1959—1963)	2014—11	28.00	377
IMO 50 年. 第 2 卷(1964—1968)	2014—11	28.00	378
IMO 50 年. 第 3 卷(1969—1973)	2014—09	28.00	379
IMO 50 年. 第 4 卷(1974—1978)	即将出版		380
IMO 50 年. 第 5 卷(1979—1984)	2015—04	38.00	381
IMO 50 年. 第 6 卷(1985—1989)	2015—04	58.00	382
IMO 50 年. 第 7 卷(1990—1994)	2016—01	48.00	383
IMO 50 年. 第 8 卷(1995—1999)	即将出版		384
IMO 50 年. 第 9 卷(2000—2004)	2015—04	58.00	385
IMO 50 年. 第 10 卷(2005—2009)	2016—01	48.00	386
历届美国大学生数学竞赛试题集. 第一卷(1938—1949)	2015—01	28.00	397
历届美国大学生数学竞赛试题集. 第二卷(1950—1959)	2015—01	28.00	398
历届美国大学生数学竞赛试题集. 第三卷(1960—1969)	2015—01	28.00	399
历届美国大学生数学竞赛试题集. 第四卷(1970—1979)	2015—01	18.00	400
历届美国大学生数学竞赛试题集. 第五卷(1980—1989)	2015—01	28.00	401
历届美国大学生数学竞赛试题集. 第六卷(1990—1999)	2015—01	28.00	402
历届美国大学生数学竞赛试题集. 第七卷(2000—2009)	2015—08	18.00	403
历届美国大学生数学竞赛试题集. 第八卷(2010—2012)	2015—01	18.00	404
新课标高考数学创新题解题诀窍:总论	2014—09	28.00	372
新课标高考数学创新题解题诀窍:必修 1～5 分册	2014—08	38.00	373
新课标高考数学创新题解题诀窍:选修 2—1,2—2,1—1,1—2分册	2014—09	38.00	374
新课标高考数学创新题解题诀窍:选修 2—3,4—4,4—5 分册	2014—09	18.00	375
全国重点大学自主招生英文数学试题全攻略:词汇卷	2015—07	48.00	410
全国重点大学自主招生英文数学试题全攻略:概念卷	2015—01	28.00	411
全国重点大学自主招生英文数学试题全攻略:文章选读卷(上)	即将出版		412
全国重点大学自主招生英文数学试题全攻略:文章选读卷(下)	即将出版		413
全国重点大学自主招生英文数学试题全攻略:试题卷	2015—07	38.00	414
全国重点大学自主招生英文数学试题全攻略:名著欣赏卷	即将出版		415
数学物理大百科全书. 第 1 卷	2016—01	418.00	508
数学物理大百科全书. 第 2 卷	2016—01	408.00	509
数学物理大百科全书. 第 3 卷	2016—01	396.00	510
数学物理大百科全书. 第 4 卷	2016—01	408.00	511
数学物理大百科全书. 第 5 卷	2016—01	368.00	512
劳埃德数学趣题大全. 题目卷. 1:英文	2016—01	18.00	516
劳埃德数学趣题大全. 题目卷. 2:英文	2016—01	18.00	517
劳埃德数学趣题大全. 题目卷. 3:英文	2016—01	18.00	518
劳埃德数学趣题大全. 题目卷. 4:英文	2016—01	18.00	519
劳埃德数学趣题大全. 题目卷. 5:英文	2016—01	18.00	520
劳埃德数学趣题大全. 答案卷:英文	2016—01	18.00	521

哈尔滨工业大学出版社刘培杰数学工作室已出版(即将出版)图书目录

书　　名	出版时间	定　价	编号
李成章教练奥数笔记.第1卷	2016－01	48.00	522
李成章教练奥数笔记.第2卷	2016－01	48.00	523
李成章教练奥数笔记.第3卷	2016－01	38.00	524
李成章教练奥数笔记.第4卷	2016－01	38.00	525
李成章教练奥数笔记.第5卷	2016－01	38.00	526
李成章教练奥数笔记.第6卷	2016－01	38.00	527
李成章教练奥数笔记.第7卷	2016－01	38.00	528
李成章教练奥数笔记.第8卷	2016－01	48.00	529
李成章教练奥数笔记.第9卷	2016－01	28.00	530
zeta函数,q-zeta函数,相伴级数与积分	2015－08	88.00	513
微分形式:理论与练习	2015－08	58.00	514
离散与微分包含的逼近和优化	2015－08	58.00	515
艾伦·图灵:他的工作与影响	2016－01	98.00	560
测度理论概率导论,第2版	2016－01	88.00	561
带有潜在故障恢复系统的半马尔柯夫模型控制	2016－01	98.00	562
数学分析原理	2016－01	88.00	563
随机偏微分方程的有效动力学	2016－01	88.00	564
图的谱半径	2016－01	58.00	565
量子机器学习中数据挖掘的量子计算方法	2016－01	98.00	566
量子物理的非常规方法	2016－01	118.00	567
运输过程的统一非局部理论:广义波尔兹曼物理动力学,第2版	2016－01	198.00	568
量子力学与经典力学之间的联系在原子、分子及电动力学系统建模中的应用	2016－01	58.00	569

联系地址:哈尔滨市南岗区复华四道街10号　哈尔滨工业大学出版社刘培杰数学工作室
网　　址:http://lpj.hit.edu.cn/
邮　　编:150006
联系电话:0451－86281378　　13904613167
E-mail:lpj1378@163.com